PRAISE FOR
RAISING PASTURED RABBITS FOR MEAT

"Nichki Carangelo's *Raising Pastured Rabbits for Meat* is like a long chat with a warm and generous friend who's sharing advice based on her own experience, hard won from both her successes and her mistakes (which you won't have to repeat). Her approach is pragmatic and flexible, never doctrinaire. Perhaps her best advice is to avoid rabbit 'monoculturing,' which is only likely to succeed at an unattainably large scale, and instead to make your rabbit marketing venture an integral part of a diverse and adaptable small farm."

—**Harvey Ussery**, author of *The Small-Scale Poultry Flock*

"*Raising Pastured Rabbits for Meat* covers some of the ups and downs of a rabbit-raising start-up, as well as the emotions and real-life mistakes tied to starting a farm. Overall it's a well thought out introduction to pasture-based rabbit production for beginners."

—**Daniel Salatin**, Polyface Farm

"They're cute, fluffy, and one of the world's most sustainable sources of meat. Yes, rabbits can make an excellent food source and income stream for a diversified family farm or homestead. Farmer and author Nichki Carangelo clearly lays out the essentials you need to start, manage, and grow a meat rabbit business in a way that not only generates income but also treats the animals humanely. There isn't another book out there like this on the subject."

—Rebecca Thistlethwaite,
author of *The New Livestock Farmer*

"As a rabbit farmer and chef who prizes rabbit as a heritage ingredient, I've been encouraged to see more farmers, cooks, and consumers rediscovering the benefits of pastured rabbit in recent years. With this book Carangelo provides a growing community of small-scale producers with an essential road map for the journey of launching a small rabbitry into a successful business and long-lasting, nourishing resource."

—**Mike Costello**, chef and farmer, Lost Creek Farm

RAISING

An All-Natural, Humane,

PASTURED

and Profitable Approach to

RABBITS

Production on a Small Scale

FOR MEAT

NICHKI CARANGELO

Chelsea Green Publishing
White River Junction, Vermont
London, UK

Editor: Makenna Goodman
Project Manager: Sarah Kovach
Copy Editor: Laura Jorstad
Proofreader: Nancy A. Crompton
Indexer: Shana Milkie
Designer: Melissa Jacobson
Page Composition: Abrah Griggs

Printed in the United States of America.
First printing November 2019.
10 9 8 7 6 5 4 3 2   22 23

Our Commitment to Green Publishing
Chelsea Green sees publishing as a tool for cultural change and ecological stewardship. We strive to align our book manufacturing practices with our editorial mission and to reduce the impact of our business enterprise in the environment. We print our books and catalogs on chlorine-free recycled paper, using vegetable-based inks whenever possible. This book may cost slightly more because it was printed on paper from responsibly managed forests and we hope you'll agree that it's worth it. *Raising Pastured Rabbits for Meat* was printed on paper supplied by Versa Press that is certified by the Forest Stewardship Council.©

Library of Congress Cataloging-in-Publication Data
Names: Carangelo, Nichki, author.
Title: Raising pastured rabbits for meat : an all-natural, humane, and profitable approach to
    production on a small scale / Nichki Carangelo.
Description: White River Junction, Vermont : Chelsea Green Publishing, 2019. | Includes
    bibliographical references and index.
Identifiers: LCCN 2019030635 (print) | LCCN 2019030636 (ebook) |
 ISBN 9781603588324 (paperback) | ISBN 9781603588331 (ebook)
Subjects: LCSH: Rabbits. | Pasture animals. | Rabbit meat.
Classification: LCC SF453 .C37 2019 (print) | LCC SF453 (ebook) | DDC 636.932/2--dc23
LC record available at https://lccn.loc.gov/2019030635
LC ebook record available at https://lccn.loc.gov/2019030636

Chelsea Green Publishing
85 North Main Street, Suite 120
White River Junction, VT 05001
(802) 295-6300
www.chelseagreen.com

FSC
www.fsc.org
MIX
Paper from
responsible sources
FSC® C005010

CONTENTS

Introduction

I owe a lot of my farming career to Polyface Farm's Joel Salatin, the author of *Pastured Poultry Profits*, or maybe to a woman named Amy who gifted me her worn-out copy of his book.[1] This was eight years ago, when I was on the first leg of what proved to be my very short training period in commercial farming. With less than one year's experience behind me—and virtually none with raising chickens—my then partner (now husband) Laszlo and I went for it and ordered 100 chicks to raise on our friend Kingsley's nearby rocky pastures.

We had no money, no markets, and no land. To be honest, we didn't have any reason to think we would ever have the resources to own our own farm. What we had instead was energy and the indestructible drive to work that I find most beginning farmers have in common. We were loaned those rocky pastures for two whole years, which provided us a place to put that energy while we got our education and found our farming team. As time passed the rest of our vision became clearer, and we slowly ground away building our diversified dreamscape one season at a time.

Starting a small, low-risk enterprise kept us safe while still giving us skin in the game. It encouraged us to make moves toward our long-term goals: to name our business, to open a new bank account, and to comb through land access listings all over the Northeast. Most important, it gave us a reason not to quit when things got hard. Never underestimate the combined power of commitment and momentum; in my humble opinion, these are the two most important factors in any start-up farm.

Our baby broiler enterprise rewarded us emotionally and —thanks to Mr. Salatin's thoughtful animal husbandry guidelines and clear enterprise budgets—financially. The success of our first batch made us excited to get another, and the success of that one made us get more still. We now raise 3,000 pastured meat birds a year—all in addition to around 30 heritage hogs, 400 laying hens, and 500 rabbits. This is a life we couldn't have even imagined back when we ordered those first hundred fluffy yellow chicks.

Salatin's *Pastured Poultry Profits* outlines a system that is realistic and functional. Most important, it makes sure to take good care of the animals *and* the farmers. It saved us countless years of trial and error, and it reliably produced the results it promised. A guide this good in farming is a treasure, and we were lucky to have it during those fragile early years when failure and frustration can be so prevalent. And while Salatin's book has been our bible for raising chickens, we didn't have a good resource to turn to on raising rabbits. Through my own experimentation and research, I am writing the book I wish I'd had when I started off.

Having officially landed on our farm six years ago—land we hope to be farming for the rest of our lives—we've been working out the kinks and fine-tuning our operation. We're not *complete* beginners, but we're early enough in our careers that the freshness of being a beginner is still there. And while we've made a ton of mistakes, thankfully we have also made many improvements, and have since developed a rabbit production system that is at once humane, profitable (for farming, anyway), and replicable. While this guide does not claim to offer information as in depth as Salatin has done for poultry, I write it with the hope that it will save you time, money, and—most important—that precious, precious energy that drives us farmers forward.

CHAPTER 1

Why Rabbits?

S trangers, when gathering, love to ask one another what they do for a living. This makes sense—Americans spend a lot of time at work, so why not start there in the search for common ground? It seems, though, that not all careers serve as good conversational fodder. Unless they, too, are in the storage industry, people usually aren't dying to know more about my pal's job at the warehouse, even though he gets to drive a forklift. But if there's a firefighter or a circus performer in the room, I'll bet you good money folks will be asking them about their work all night long. Telling people you're a farmer, at least in my neck of the woods, can beget the same reaction.

I'm not exactly sure why a job in agriculture piques people's interest so much, but I have a theory. Historically, America was an agricultural nation, and despite the decline of the agrarian lifestyle thanks to an increasing trend toward urbanization, a lot of us still have at least a distant connection to or a fond memory of an old neighborhood farm or family ranch. I think these nostalgic ties are what spark this special sentiment toward farming in the American cultural fabric. And above all that, everyone eats, which means we're all deeply connected to agriculture whether we recognize it or not.

When I'm in a new place and the inevitable "What do you do?" question finally pops up, I ready myself for the series of questions

A few young rabbits on our farm.

usually headed my way. While it can get boring saying the same thing over and over again, I'm honored to do work that people care about. As such, it's easy to answer honestly and simply.

"Did you grow up on a farm?"
No, actually, I grew up in a small city.

"How did you get into farming, then?"
I ask myself that same question all the time!

"How early do you wake up in the morning?"
Definitely not as early as you think I do. Unless it's a market day, in which case probably way earlier than you think I do!

"What do you grow on your farm?"
I answer this last one the same way every time, for the sake of expediency: vegetables, greens, herbs, flowers,

pork, chicken, eggs, and rabbits. If I say it out of order, I can't even remember all the things we produce on our farm. Even though pigs are my favorite animals to talk about simply because they're such smart and entertaining creatures, people usually gravitate toward the rabbits, because they either (a) loved eating rabbit as a child, or (b) don't know how I could eat something so cute. Either way, the conversation will sire another question: "Why rabbits?" There are two ways for me to answer this question. First, why I personally raise rabbits, and second, why farmers in general should raise rabbits.

Why I Raise Rabbits

I didn't grow up on a farm. I was, however, raised in a family full of super-resourceful immigrants who had grown up on farms. On both sides of my family, I had grandparents with virtual oases for backyards. I'm talking stucco sheds covered in grapevines, rows of peach and pear trees, and tomatoes, eggplants, and peppers stuffed into every break in the asphalt. The homes of my relatives and their friends were total anachronisms in their urban and suburban Connecticut neighborhoods—little Old World islands floating in a sea of Americana.

Tucked into of one of these mini homesteads, behind a garage kitchen used pretty much exclusively for frying stuff (a nifty feature I have yet to see outside Italian American households), was Maria and Pasquale's rabbit barn. In true big-family fashion, I don't really know what my relation to Maria and Pasquale is. Maybe they're my grandfather's cousins or maybe they're just some other expats from Benevento who became chosen family to the Carangelos in the New World (these nebulous family ties are common where I'm from—once I went 15 years thinking someone was my aunt Anette, only to find out she was actually a distant relative named Antoinette). Tiny yet tidy, Maria and Pasquale's urban rabbitry is still the sole provider of fresh rabbit meat in the entire city of Waterbury, Connecticut—where I grew up. Pasquale, even in his 80s, can kill and clean a rabbit with nothing but a sharp knife. I spend a lot of time in slaughterhouses

My great-grandmother, aka Nonni, at her backyard basil patch in Waterbury, Connecticut. *Photograph courtesy of Sandra Kelly.*

these days and have processed at least a couple thousand animals myself, and I still have never seen anyone as quick and as clean as he is. I used to joke that Pasquale could kill and clean a rabbit in a tuxedo and not even need to wash his hands before heading to a wedding. He's honestly that good.

So when Laszlo and I wanted to start raising our own food but didn't have a big farm on which to put livestock, we simply followed in the footsteps of my family members and their old-fashioned friends. We went and got ourselves some rabbits, and we've held on to them ever since. And while we started raising rabbits mostly on a whim, we were lucky that they happened to measure up when things like unit economics, market potential, and environmental impact became really important factors in our decision making years later. In fact, we found there are many good reasons to raise rabbits either commercially or as a hobby.

Why You Should Raise Rabbits

A 2010 article published by *Good* magazine featured locavore trend-setter Michael Pollan making a big claim: "Rabbits make more sense than chicken."[1] In a country with a $41 billion broiler chicken industry, Pollan's statement seems pretty bold, but *Time* magazine agreed in its piece "How Rabbits Can Save the World."[2] Even the *New York Times* has hopped aboard the bandwagon.[3] According to recent reports from gourmet food supplier D'Artagnan, rabbit sales have been increasing over the course of the last decade.[4] They even report their own rabbit sales doubling over a four-year period. The wider world is abuzz about the supposed super protein, and yet very few farmers are stepping up to meet the rapidly increasing interest in sustainably raised rabbit. I can't help but wonder why, as I see tremendous opportunity for any new and seasoned agrarians willing to learn this increasingly lost art of raising such triple-purpose wonders. With comparatively little money, space, or labor, rabbits can earn you an honest living and return some much-needed (and demonstrably desirable) diversity back into the markets for food, fiber, and fertility.

For starters, rabbits are easy to handle. Unlike a lot of other livestock, they're small and docile. Well-bred rabbits from good stock are unaggressive and easy to pick up and move around. Plus, they aren't heavy. Even the largest breed of domestic rabbit, the Flemish Giant, maxes out at a very manageable 22 pounds (10 kg).[5]

Rabbits, and everything they require, are light and portable. This is true in terms of both daily chores and major moves, like relocating to a new piece of land. Aside from mobile rabbit tractors, of which you'll need some version, should you decide to use a pasture-based method, the heaviest item on the list is going to be a wheelbarrow. Even the wire-pasture rabbit tractors we use are light, at least when compared with those we use for our meat birds—they're only half the size in terms of area (6 × 9 feet versus 9 × 12 [1.8 × 2.7 m versus 2.7 × 3.7]).

The fact that rabbits and their associated tools and equipment are not heavy is especially important for folks with physical limitations. That rabbits are so portable makes them a great starter enterprise for farmers and homesteaders without secure land tenure. Unlike some larger livestock, like dairy cows, who require more permanent infrastructure like a milking parlor, or goats who may need extensive fencing systems, everything you need for a rabbitry can fit right into a U-Haul and be moved across town or even cross-country in a jiff. If your journey to landownership is anything like ours was, this feature is guaranteed to matter. Before landing on our "forever" farm in Hudson, New York, Laszlo and I moved our animals four times in three years.

Another quality that makes rabbits a solid starter enterprise is that they have a low barrier to entry. Rabbits are small and therefore need comparatively little space, even in pasture-based systems. Beyond that, rabbits require minimal start-up capital. In fact, a well-managed commercial rabbitry can repay its entire initial investment in a single year. Even small-scale hobby rabbitries can start saving homesteaders money within a couple of months. Later on in this guide, you'll find an enterprise budget that breaks it all down in detail, but for now just know that you can start raising rabbits in your backyard for less than $500.

Another nice thing about rabbits is that they are small enough to be transported in containers you probably already have, especially if you're interested in raising only a few of them. Don't feel like spending $80 on a poultry transport crate? Your cat carrier will do just fine. Is it too early to invest in a pickup truck or a livestock

RABBITS AND WORMS

The benefits of a wormery in your rabbit barn are twofold: It can save you money on amendments and fertilizers, and it can also help you earn extra income selling worms to folks who fish, have worm-eating pets, or use compost bins of their own. Farmers can even sell finished *vermicompost*, or worm manure, to gardeners. Happy D Ranch Wormery has a great in-depth resource on their website for how to start a worm farm in your rabbitry, but the basic principles are pretty simple.[6]

The first step is to build a wooden frame on the floor beneath your rabbit cages. Happy D Ranch recommends the walls of this frame be 12 inches (30 cm) tall and 5 inches (13 cm) wider than your rabbit housing in all directions. The purpose of this frame is to contain the rabbit pellets as they fall through the wire and onto the ground. Once your frame is in place, add some carbonaceous material, such as dry leaves, straw, or hay, to form a layer 2 or 3 inches (5.1–7.6 cm) deep. When the frame and the bedding are in place, it's time to wait for your rabbits to poop. Once you've got a couple of inches of manure piled onto the bedding, use a shovel to mix it together with the bedding and then soak the mixture thoroughly with a hose. This moisture kick-starts the decomposition process, which eventually causes the mixture to heat up. Turn your mixture once per day, checking each time to see if it's still warm. Once the mixture is cool to the touch, it should be ready to host your worms!

The recommended density for worms in this system is 300 to 500 per surface area square foot (0.1 m²). You can order your worms online or purchase them from a local worm farm if there's one near you. At the time of writing, the cost for 1,000 red wiggler worms, which are great for compost systems like this, is about $25. Once released, the worms will disappear into the rabbit manure and start turning those round little "bunny berries" into a rich, nutrient-dense humus—a stellar soil amendment. We don't currently have a worm farm in our rabbitry, but adding one is on our to-do short list. It's an incredibly low-cost investment that can help maximize both ecologic and economic value per square foot on our farm—plus, Laszlo loves to fish.

trailer? Almost everything you need for rabbits will fit right into the back of your sedan.

In particular rabbits end up saving you a boatload of money (and time!) when it comes time to process them. Rabbits don't have feathers, and this single characteristic is what makes them much easier to slaughter and dress than most of the other small livestock out there. For comparison, to efficiently process chickens you need a whole fleet of equipment: kill cones to hold the birds in place, a scalder to loosen up the feathers, and a mechanical plucker to pull them all out. Just these three items alone can run you a couple thousand dollars. Rabbits, on the other hand, require as little as a single sharp knife (more on butchering methods in chapter 13). Since they're typically sold whole, rabbits are easy to package and store, and readily available shrink-wrap bags make fresh rabbits freezer-ready in seconds.

Lastly, rabbits are an asset to the diversified farmscape. They can be tucked into the back of existing barns and sheds and grazed on small patches of marginal land, even ahead of your chickens or behind your ruminants on lush pastures. Their presence in fields and yards is beneficial for a host of reasons. For one thing, rabbit manure is incredibly nutrient-dense. It's full of nitrogen, phosphorus, potassium, minerals, and a whole host of micronutrients. According to the Michigan State University Extension, it has four times as many nutrients as cow or horse manure and twice as many as chicken manure.[7] Even better, unlike that from the omnivores on our farm (chickens and pigs), manure from our herbivorous rabbits is "cool," as opposed to "hot." This means rabbit poop innately has the ideal carbon-to-nitrogen ratio (about 25:1) and therefore does not need to be rotted with leaves or other carbonaceous material before use on plants. Hot manures, on the other hand, have too much nitrogen for each part carbon and, as a result, will burn plants if not properly amended and composted beforehand. (See "Integrating Rabbits and Vegetables at Letterbox Farm" on page 14 for more information on the benefits of rabbit manure.)

And it gets better. Since rabbit manure is organic matter, it works to build soil structure. Good soil structure means better drainage and moisture retention, both of which are becoming more and more important considering our increasingly unpredictable climate.

For these reasons, worms especially love rabbit manure, and many rabbitries double as wormeries. Simply put: Rabbits produce a lot of manure. One doe and her offspring can produce an entire ton of manure in a single year—and beyond adding nutrients for your own farm, this could be marketed and sold as a value-added product.

Rabbits are also a great source of nutrition. They are a lean protein, which makes them even lower in fat than chicken, their more popular culinary cousin. They are also super high in protein. Per 85 g of meat, rabbits have whopping 28 grams of protein. Compare this with pork (23 grams), beef (22 grams), chicken (21 grams), lamb (21 grams), bison (22 grams), and the closest competitor, turkey (24 grams)—and rabbit wins by a long shot.[8] Rabbit is a great source of iron, with 4 mg per serving. Rabbit meat also has fewer calories per serving than most of these meats, with turkey being the only exception.

While I believe that all livestock can have positive and restorative environmental impacts when thoughtfully raised, rabbits are especially ecological. In fact, they have a number of attributes that are catching the attention of environmentalists and food security researchers on the world stage today. The contemporary progression of global warming trends has resulted in a host of climate-based impacts that have had very real effects on farmers everywhere. Extended drought and the shrinking availability of arable land has many looking for livestock with lower carbon footprints. Rabbits shine in this regard because they can be raised on marginal land and are very efficient converters of food and water. This means we need less land, less water, and less energy to grow feed for rabbits than most other livestock. Rabbits convert food and water into edible meat 1.4 times more efficiently than pigs and 4 times more efficiently than sheep and cattle.[9]

Thankfully, these benefits don't come at the expense of flavor. Unlike some other foods that rank high in sustainability but low in palatability (looking at you, Japanese knotweed), rabbit meat is versatile and delicious. Meat from young rabbits is mild, sweet, and tender. It can be paired with a wide range of flavors and takes nicely to almost any preparation used for chicken. A rabbit is simple to break down and can be roasted, fried, braised, grilled, or poached. It can even be preserved in cured sausages and rillettes.

HOW TO RUN A SUCCESSFUL RABBITRY

While there are a lot of reasons you should be raising rabbits right now, rabbits come with their unique set of challenges. It is important to consider each one before making any investments. Like most other agricultural enterprises, a rabbitry has slim margins, so the success of your operation is directly related to your ability to manage several important aspects. The most important of these is optimizing animal health and well-being.

Rabbits are particularly sensitive to disease and illness, which is precisely why they are not commonly raised in large-scale, commercial settings. These delicate creatures cannot handle the stress of being raised in poor conditions, so it's imperative you keep them clean, dry, and comfortable at all times. Of course, even in the best conditions rabbits can get sick, but good management and wise breeding can eliminate most issues over time. (We will dive deeper into rabbit wellness in chapter 9.)

The second key to running a successful rabbitry is achieving uniform and reliable production. This requires a bit more savvy than in some other livestock enterprises. Unlike the case of poultry, where farmers can buy day-old chicks, or pork, where you can get feeder pigs that are already eight weeks old and much less vulnerable, rabbit farmers have to do their own breeding. Poor or inconsistent litter sizes or growth rates can tank an operation, so developing an eye for good genetics and keeping to a consistent schedule are paramount.

In addition to efficient production, you will need steady sales if you are running a commercial rabbitry. Lean margins mean a successful rabbitry will rely on optimal production and strong sales at the right price. Keeping costs low and efficiency high probably seems like a no-brainer, but we will see later on how even small shifts in production or sales can make a big difference for your bottom line.

Lastly, commercial rabbit farmers need access to a good, legal, and affordable processor. While backyard producers are allowed to slaughter their own livestock, the rules for processing salable meat can be very different, depending on where you are. Some states, like Vermont and Maine, allow producers to slaughter on their own farms. Others like New York and Rhode Island have their own state-inspected facilities, while others allow rabbits for sale to be slaughtered only at a US Department of Agriculture (USDA)–inspected facility. It's important to know your local rules and regulations before getting your rabbits.

Do not let these challenges discourage you. The purpose of this guide is to walk beginning farmers (or those already farming and new to rabbits) through each step while demonstrating how we at Letterbox Farm have tweaked our production systems and marketing relationships over the years in order to grow a successful rabbit enterprise. I hope our countless trials, subsequent errors, and recalculations and adaptations will save you time, money, and precious energy as you begin or expand your rabbitry.

Finally, rabbit adds diversity to your diet. The traditional American dinner plate had a lot more than beef, chicken, and pork on it. Prior to the widespread implementation of *CAFOs*, or concentrated animal feeding operations, which heavily focused on the production of these three proteins, our dinner tables were laden with a wide variety of proteins, like venison, bison, pheasant, and of course rabbit. To eat rabbit is to eat food that is rich in heritage.

Integrating Rabbits and Vegetables at Letterbox Farm

By Faith Gilbert, Vegetable Manager

While this book focuses on our livestock operation, our farm is actually split revenue-wise 40–60 between raising livestock and growing produce in our 2½-acre (1 ha) market garden. Managing fertility is an ongoing task in growing vegetables, and producing fertilizer for vegetable production was one of our original draws to raising rabbits. We imagined rabbits as the *ideal* small-scale fertility source: light, easy-to-contain animals that produce an abundance of rich manure that's easy to spread. In fact, even before we brought together our diversified farm, I had drafted rabbits into my first-ever crop plan, imagining them grazing down the clover access paths on a ½-acre (0.2 ha) vegetable market garden. Our plans have evolved quite a bit since then, as has our scale of vegetable production, but I still think rabbits are the perfect addition to vegetable operations at any scale, including (and especially!) very small farms and community gardens that want to produce fertility on-site.

We practice a form of vegetable, herb, and flower production known as *small-scale intensive*, meaning we aim for high productivity on a limited growing space. By double-cropping where possible, planting densely, and turning over and replanting beds quickly, we squeeze 4 acres (1.6 ha) or so of crops into 2½ acres (1 ha) of beds, resulting in higher revenue per acre. There are pros and cons to this method, but it works well for our personal goals and with our landscape.

Managing soil fertility is key in any produce operation, but especially in small-scale intensive growing, where fertility needs are higher and the ability to grow fertility through cover crops is more limited. Most small-scale growers rely on bringing in composted animal manures to maintain nutrient levels. Rabbits and their manure possess some unique qualities that make them an excellent companion to small-scale intensive vegetable production:

Pound for pound, rabbit manure is richer in nutrients than cow, horse, chicken, or any other common farm animal manure by a significant margin. More nutrients per pound means less labor hauling it around, especially if you're hauling and spreading by hand.

Unlike other manures, it can be directly applied to crops without burning them, meaning you can skip the labor and space for composting it (at least from the plant's perspective—food safety concerns are another matter).

While the actual nutrient composition of an animal's manure will vary, most sources benchmark rabbit manure's composition as about 2.5 percent nitrogen, 1.5 percent phosphorus, and 0.5 percent potassium (2.5–1.5–0.5). In addition to these macronutrients, rabbit manure contains organic matter, trace nutrients, and microorganisms, all contributing factors that build soil structure, plant health, and a vibrant soil food web.

Compared with other animals, rabbit manure has a higher ratio of nitrogen relative to phosphorus and potassium. Many common crops, like brassicas and baby greens, have a light need for phosphorus and potassium (P and K), but a high need for nitrogen (N). Additionally, while P and K are relatively stable in the soil, nitrogen is more soluble and needs to be applied regularly. Therefore, farms that use animal manure compost as their nitrogen source often end up regularly overapplying those other nutrients, which can have ill effects on long-term soil health, especially growing in hoop houses or in low-rainfall areas. Rabbit manure's higher nitrogen ratio, then, makes it easier to avoid overamending nutrients that aren't needed.

Application Method and Rate

One of the other key benefits of rabbit manure is its ease of application. It can be directly applied to crops without "burning" them, unlike other animal manures. It is pre-pelletized and easy to spread by hand (if needed), and is one of the drier and less smelly manures (if collected in a well-drained spot).

We most often amend for nitrogen, rather than other macronutrients, because our soils are well balanced for most other nutrients and hold fertility well. The nitrogen recommendations for most vegetable crops range between 50 pounds per acre (56 kg per ha), for light feeders, and 100-plus pounds per acre (112-plus kg per ha) for heavy feeders. With rabbit manure's composition of 2.5 percent nitrogen, that's 2,000 pounds (900 kg) of manure per acre for light feeders, double for heavy feeders. Translated into per-bed applications, that's about 14 pounds (6.4 kg) for light feeders and 28 pounds (12.7 kg) for heavy feeders per 3 × 100-foot (1 × 30 m) garden bed. On a per-bed basis, that's a pretty attainable rate of one or two 5-gallon (20 L) buckets of fresh manure for all your nitrogen needs. For the first few years of our operation, that's just what we did: load some buckets of manure into a garden cart and haul them up to the beds we were preparing to plant.

It worked pretty well, although the task proved burdensome when prepping multiple beds at once. As we scaled our vegetable garden from ½ acre (0.2 ha) to 2½ acres (1 ha), fertilizing beds one by one became impractical. A few years in, we invested in an ATV and a small trailer, which we now use for all manner of livestock and vegetable tasks. This is a step up, allowing us to haul much more weight and volume than is practical in a garden cart. However, it still feels inefficient for our scale, especially when compared with the speed of applying granulated chicken manure or other dry fertilizer. Looking forward, we're excited about the potential for an ATV-pulled manure spreader, and planning to muck out the rabbit house right into the spreader to distribute over the fields in the fall.

As a final note on application, while rabbit manure's pelleted texture makes it arguably the easiest manure to spread, it's still too cloddy to apply directly onto a fine seedbed without getting in the way of seeding. If spreading manure right before seeding is an important part of your vegetable system (and works with your food safety plan), it would be worthwhile to compost it first into a finer texture.

To calculate the amount of manure needed, take the desired nitrogen application rate (usually expressed in pounds per acre, available from extension websites and soil test lab recommendations). Divide the desired N by the percent N of rabbit manure (varies, but benchmark at 2.5 percent). To calculate a per-bed application, first find the square footage (m²) of your bed dimensions, and divide that by the square feet in an acre (bed square footage / 43,560 [or the m² in a hectare: bed m² / 10,000]). Now multiply the pounds of manure per acre by that fraction (or multiply the kg of manure per hectare by the metric result).

Considering Food Safety

Farmers pursuing organic certification, those who choose to follow Good Agricultural Practices (GAP) standards, or those who'll be regulated by the recent Food Safety Modernization Act (FSMA) will need to follow protocols for safe application of manure. Both National Organic Program standards and GAP protocols limit the application of raw manure to no fewer than 120 days before harvest if it's a crop that touches the ground, and 90 days for crops that don't. In addition GAP requires waiting two weeks after applying manure before planting any crop. FSMA rules are currently in limbo as the FDA works to determine a minimum interval between application and harvest. Currently FSMA regulations do not apply to farms that gross under $500,000 in revenue and that sell the majority of their products within 275 miles (443 km)—although that may change in the future.

Considering Financial Value

Integrating rabbits with growing produce also brings financial benefit. In times when we don't use manure from our own animals, we spend money managing fertility in other ways, whether on organic fertilizers or finished compost, amounting to $500 to $1,000 per year total for our vegetable operation. While not all rabbit raisers also run vegetable farms, there may be demand from gardeners and vegetable producers in the area. Some rabbit caretakers are able to offset the cost of raising their animals or augment their meat rabbit income by selling buckets, bags, or "shovel your own" rabbit manure. Sites like Craigslist, Amazon, eBay, and Etsy show backyard and commercial rabbit operations selling manure in many forms, for prices ranging from $1 to $14 per pound.

CHAPTER 2

Rabbits on the Farm

Letterbox has been a diversified farm since its inception, so it was relatively easy to fold rabbits into our existing production plans and sales outlets. To give you a better idea of how rabbits fit into the broader landscape of our farm, this chapter will provide a summary of our business and how it works. It has been crucial for us as we develop our own systems to see what models out there work, in order to understand where our model fits in.

Our Land and Community

Letterbox Farm sits on 64 acres (25.9 ha) of fields, woodlot, and ravines in New York's beautiful, bountiful Hudson Valley. In the past this land served as a dairy, an orchard, and then a hayfield. We know this because the remnants of each of these past incarnations can be seen today lurking inside our barns, up in the rafters of the shop, and in the piles of old, rusty equipment we inherited with our fields. We purchased our land when the elderly farmer from whom we leased it for two seasons passed away with no heirs. In order to help fund our purchase, we sold the development rights to 56 of our 64 acres to a land trust. This decision had two major benefits: It brought down our purchase price by a third, and it guaranteed that our farmland would be preserved for generations to come.

Our community is home to roughly 7,000 year-round residents of diverse cultural and economic backgrounds. Our farm sits just on the edge of our dense little city, which sort of makes us Hudson's backyard. We are 200 miles (322 km) from New York City, which is just close enough for a taste of the benefits of the seemingly bottomless marketplace but certainly not close enough to rely on it.

We are fortunate to share a county with hundreds of other farmers, both beginning and seasoned. This unusually high concentration of agrarians gives us access to a valuable network of resources and knowledge. Being a part of such a community allows us to swap seedlings when our kale doesn't germinate and borrow expensive poultry crates by the truckload when we are picking up pullets from Pennsylvania. A robust Google group organized by the Hudson Valley Young Farmers Coalition allows us to find advice on local slaughterhouses and where to source the cheapest tomato boxes. I can't tell you how many times we have been bailed out by the generosity and intelligence of the citizens of our little town. There are just way too many to count.

Letterbox Farm is co-owned by three full-time, year-round farmers, each of whom is the lead on specific tasks and enterprises.

We have a vegetable manager, Faith, who is in charge of planning and executing the many moving parts of our growing market garden. With the integral help of our seasonal crew members, she grows 2.5 acres (1 ha) of intensively managed greens, herbs, vegetables, and edible flowers.

Our big-picture land steward and mechanical genius is Laszlo, who takes care of our farm's day-to-day maintenance needs, like mowing the lawns and brush-hogging the fields while also keeping our tractors, vehicles, and other equipment in top condition.

And then there's me: our livestock manager. With the help of a dedicated livestock crew member, I currently steward our roughly 400 (it all depends on the hawks, really) free-range laying hens, 30 or so pastured hogs, and a 24-doe rabbitry, as well as managing our pastured poultry, which at 3,000 meat birds per year is the backbone of our livestock operation.

In addition to the three core farmers, we bring on four full-time crew members each year and a couple of outstanding volunteers to

Moo, my most constant companion.

help us keep the farm running smoothly. All of our full-time staff (including Laszlo, Faith, and myself) derive their primary income from farming at Letterbox. Although some of us occasionally take on side work, most often in something agriculture-adjacent, like farmer education or research, we very much rely on our business to cover all of its own expenses (including the mortgage and all the other capital investments) and pay our salaries.

While many farms depend on various forms of unpaid labor, like apprentices, interns, and work traders, our operating budget is built on the assumption that every necessary task is being performed by staff that's paid and on the books. We're blessed in that our one regular volunteer, Eden, is insanely reliable, dedicated, and hardworking, but in general we didn't think it wise to build a business plan that depended on unpaid labor, even though I very much understand why it's common practice in the low-margin, high-risk world of agriculture.

As you design your rabbitry, consider your own expenses and labor model and plan to scale accordingly. Do you, like us, need the farm to pay your mortgage? Are you trying to support yourself farming, or is your rabbitry a side job for extra cash? Will it just be you, or will you have a team? The more realistically you answer these questions in the beginning, the more accurate your business plan will be when the rubber actually hits the road.

Get to Know Your Markets

The best thing about running a diversified farm is that it can unite a range of strengths, weaknesses, personalities, and bodies while working toward a common goal. We love the breadth of activities we host on our farm—especially when it comes time to make dinner—but producing such a variety of products can be a real logistical nightmare when it comes to some tasks. Take building markets, for example. A great outlet for edible flowers isn't usually an ideal market for pork chops or cabbage, so we have to work extra hard to make sure we have a home for every product.

In our search to find a lid for each of our many pots, we've had to build *a lot* of different kinds of markets. So many, in fact, that when folks ask me how we sell our food, I often contemplate whether it'd just be faster to just list the ways we *don't* sell our food. Right now, we sell six days a week (seven if you count our humble farm store) through three main outlets.

Full-Diet CSA

For those who don't already know, *CSA* stands for "community-supported agriculture," which is basically a subscription service for farms. While the concept has been steadily growing and evolving since its US debut in the 1980s, in the traditional model community members pay a farm a set fee in advance and then receive a share of what the farm produces each week for a certain number of weeks. In our full-diet version, members take home a dozen eggs and a pre-selected share of five to seven vegetables each week in addition to whatever meat they choose from our fridge or freezers, which are stocked with a variety of cuts of chicken, pork, and rabbit. The value of the meat each member chooses is then deducted from a credit that the member begins with at the start of their membership. This allows folks to eat the proteins they like best while also making sure our farm is compensated fairly for the product.

We run our 60-member, full-diet CSA in three waves, which include a 6-week spring share, a 20-week summer share, and an 8-week fall share. Because most folks who join a CSA do so as a means of replacing their weekly trips to the grocery store, we aim to fill our shares with familiar crowd pleasers and household staples, like tomatoes and onions, rather than some of our more obscure crops like chicories and chervil. As far as meat is concerned, our members often opt for easy, quick-cooking items like sausages and chicken breasts, reserving less familiar items like rabbit for special occasions. While demand can vary depending on season, contemporary culinary trends, and holidays, only one or two of our CSA members will choose to take home a rabbit in an average week.

Farmers Markets

Thursday afternoons we spend on the farm managing our CSA pickup, but Saturday and Sunday we go to market. Currently Letterbox is in three farmers markets. We're fortunate in that our markets are, for the most part, well trafficked and profitable, but this isn't always the case. Be sure to do your research before jumping into the first open market stall you can find. In my secondary role as our farm's manager of all things retail, I look for a few specific things when considering new markets:

What other farms are there? I like to be in good, professional company at my markets, so I look for reputable, complementary vendors. The complementary part is key—seek out markets that have a good mix of businesses but aren't oversaturated with any of the products you'll be selling. It's unlikely you'll come across another rabbit vendor as you scout potential farmers markets, but if you find one, make sure it's busy enough to support you both before you join.

How well is the market managed? I look for markets with well-maintained websites and frequently updated social media accounts. These are good indicators that the market manager is taking the job of driving traffic seriously. If you can, avoid markets that have a high turnover of vendors. If businesses keep leaving, it means they're not happy with how things are going.

Is it the right size for us? While we love selling out, it's important we have enough product at every market to make a good showing for most of the market day. Unless you're going to grow into them within a season or two, avoid joining markets that are too high-volume for you. I offer this advice because if you're always sold out when customers stop by your stand, they may stop coming altogether.

Another thing to consider when choosing a market is where it is in proximity to your farm. The major downside to being a market farm, at least in our case, is the travel. Our Sunday market is only a couple of towns over, but our Saturday markets are almost 100 miles (160 km) south of our farm. Markets opening at 8:30 AM have us up at 3:30 AM, packing the van at 4:00 AM, and departing sharply at 4:15 AM. Breaking down tents, tables, and signs plus a dash of afternoon traffic often means we are not back on the farm packing out until 12 or 13 hours later. Plus, markets can be expensive—booth fees, gas, tolls, staffing, and ice can really add up.

The upside to the farmers market is that you can bring products in whatever volume you have them, which makes them a nice outlet for a start-up farm still working out some kinks in production. Unlike our CSA, where we need very specific quantities of particular items, markets are more flexible. It's okay if we only have 57 quarts of shishito peppers and 23 bunches of fennel. Farmers markets also allow us to connect with a much larger audience than our CSA does. On any given weekend at least a few thousand people will walk by our stands, see our signage, sample our products, or grab a business card. When it comes to selling something uncommon like rabbit, exposure can be everything. It's important for your customer to know where to find you, and farmers markets provide both visibility and fodder for word of mouth. Not to mention the fact that a well-run farmers market will have a healthy presence online, further aiding your connection to consumers.

Restaurants

CSA and markets make up the bulk of our retail business, but about 50 percent of our income comes from wholesale accounts. On Tuesdays and Thursdays we are out delivering food to our local restaurant accounts. To date we work regularly with 10 or so chefs who order from us pretty much every week during the season. It is inside these relationships that our more specialty crops really shine. Edible flowers, bronze fennel, and red-veined dandelion often aren't great CSA or market offerings, but in the right hands they can transform an ordinary dinner into something stunning. On Mondays we're usually packing out for a handful of local aggregators, which are basically middlemen-type companies that buy food from Hudson Valley farms and distribute it in a variety of ways throughout the state. While these types of accounts typically garner the lowest prices for our products, they require the least amount of work (and money) on the sales end. After all, there's no labor, no gas, no ice, and no tolls.

On-Farm Sales

When I first envisioned the market for our pastured rabbit meat, I was convinced that the bulk of our product would be going to fancy restaurants down in New York City. While I wasn't totally wrong (we do sell a fair amount of rabbit to NYC restaurants), I definitely underestimated another major outlet: random drive-by traffic. Although our farm is in a rural community of only 7,000 residents, we are located on a main road, which allows for several thousand passersby to see what we're up to each day. I had hardly recognized this as an asset until one season when I moved our rabbit tractors up from the back fields to the lawn by the road. It wasn't long before strangers came knocking on our door, inquiring if they could buy some rabbits. They were thrilled to have found us.

At first, I wondered why it had taken us so long to connect with these eager patrons. We have a CSA, we're in a bunch of farmers markets, and we're on the internet, for heaven's sake. Surely people who were interested in local food would have no problem finding Letterbox Farm! What I realize now is that these newfound rabbit enthusiasts weren't necessarily interested in the sustainable agriculture movement at all. I had overestimated the popularity of rabbits with the foodie crowds and grossly underestimated the protein's resonance with folks from other countries and different generations and among traditional homesteaders, many of whom were running in different circles than we were. This oversight is funny to me now, especially since my inspiration for starting a rabbitry arose from watching my own older immigrant relatives raise rabbits in their backyards. They definitely did not do it because it was trendy.

While many Americans are only now trying rabbit for the first time, it's a staple in other parts of the world and part of a traditional diet for many here at home. In short, there are plenty of people out there who want rabbit. They're just not necessarily at the farmers market or searching on the internet. Many of them, like my grandma for example, are simply scanning the roadway on their way to Kmart. Putting a little sign out there for them to see could go a surprisingly long way. That season when we moved our rabbits up by the road, we picked up three different accounts, each of which was a large family who had been looking for local rabbits for a long time. To date, we sell between 50 and 75 rabbits a year to these families alone.

Our Farm Priorities

Even though Faith, Laszlo, and I have wildly different personalities, skills, and interests, we all work together toward a common goal. On our farm, we unanimously prioritize three things:

1. Land stewardship and animal welfare
2. The production of the highest-quality products sold at appropriate prices
3. The health and well-being of our community and our farmers

In regard to this last point, our target goals for ourselves and every member on our staff include an average workweek of 40 hours, paid time off and sick leave, and a civil servant's salary. It is also important that we create work that is meaningful, engaging, and reasonable as we grow, evolve, and age.

Five years into the wild ride of running Letterbox Farm, we have not yet managed to meet all of these goals. Each year is better than the last, though, and we have made tremendous progress over the course of what feels bizarrely like both the longest and the shortest years of my life. Much of this progress is due to the fact that now, we vet every farm enterprise we have for its ability to contribute to those goals. If an enterprise doesn't move the needle in the right direction, we reevaluate the idea to see if we can pursue it in another way that does. If we can't find a new path for an idea, we'll put it on the shelf to see if it makes more sense later, under different circumstances. This wasn't always the case, and we definitely spent a few years throwing spaghetti at the wall just trying to see what would stick. Our micro goat herd and our fledgling farm soda company, along with our beautifully speckled quail eggs (that no one wanted to buy), are in the "wrong place, wrong time" folder. But not rabbits. Thanks to their low initial investment, minimal space requirements, and comparatively nominal labor inputs, rabbits fit into our plan right away. This coupled with the fact that we've sold every rabbit we've ever grown (and for a fair price!) means they are here to stay.

CHAPTER 3

Choosing Your
Production Method

W e consider ourselves incredibly fortunate to share the
Hudson Valley with dozens upon dozens of top-notch
producers. We are even luckier to have access to a host
of local and national information-sharing networks and platforms.
Farming in the age of the internet and social media affords us exposure to innovative ideas at lightning speed, and I cannot overstate
how much our farm has benefited from experiences that others have
generously shared.

However, information overload is a real thing, and having the
opportunity to constantly compare your farm with others can easily
make you second-guess yourself. That's why it is imperative to remember that every farm is its own fickle ecosystem, each one bound by its
specific restraints and privy to unique opportunities. If there is one
thing we have learned over the course of our collective 15-year tenure
at Letterbox, it's that *there is no one right way to farm.*

If you had hoped upon picking up this book that I was going to tell
you exactly how you should raise your rabbits, brace yourself—I am
not. Just like chickens, tomatoes, parsnips, and sheep, rabbits can be
raised using a whole spectrum of methods. By the end of this chapter,

though, we will have taken a detailed look into the system we use at Letterbox, and how it might work for you, or how you might adapt it to. Which system you ultimately choose will depend upon your own specific goals, skills, limitations, and resources.

The same is true when it comes to scale. Determining the size at which running a rabbitry is worthwhile depends on two things: your farm's overhead expenses and how much money you want to make. Overhead expenses include all the expenses necessary to operate a business that cannot be conveniently traced back to any particular enterprise. This includes things like marketing costs, accounting fees, advertising, insurance, rent, utilities, web hosting, vehicle maintenance, and office supplies. The more your rabbit enterprise can piggyback on existing costs, the more profitable it will be. The more investments you need to make in order to store, market, or sell just your rabbits, the less profitable it will be. As you read this book, consider your own unique circumstances and personal ambitions, then make adjustments accordingly.

Popular Frameworks for Raising Rabbits

Throughout my own farm tenancy, I have witnessed rabbit raising techniques that range from the orthodox, like raising a couple of bunnies in backyard hutches, all the way to the downright zany. There is one example of a system I encountered that I think illustrates the breadth of possibilities particularly well.

While at the slaughterhouse one summer day, I struck up a conversation with a gentleman who ran a nearby hay farm. When he saw I was there to process rabbits, he told me he'd recently inherited a breeding pair of his own. As we chatted, I learned that instead of just getting a couple of hutches to keep them in—as most people would opt to do—he let them loose into the loft of his barn. While he did give them water, he decided to leave well enough alone beyond that. The doe and the buck ate whatever they wanted of the hay he stored there and procreated on their own accord. Before he knew it, this guy had

RABBIT TERMS

Every field has its own industry-specific jargon, and the rabbit raising world is no different. If you're new to these animals, these rabbit terms should catch you right up to speed. While there are certainly many other terms you should know when raising rabbits, to make it simpler I'll define them one by one as they come up.

Doe. A female rabbit.

Buck. A male rabbit.

Junior or grow-out. A young rabbit that hasn't reached maturity.

Kit or kitten. A baby rabbit.

Weanling. A six- to eight-week-old rabbit that no longer relies on milk and can be separated from its mother.

Kindle. Give birth.

Dry doe. A doe that is not lactating.

Breeders. Rabbits used for mating.

Herd. A group of rabbits.

Litter. A group of rabbit kits.

Fryer. A young, tender rabbit that is under 16 weeks and less than 3½ pounds (1.6 kg) dressed.

Roaster. A rabbit with firmer meat that is older than 16 weeks and weighs between 3½ and 5 pounds (1.6–2.3 kg) dressed.

Stewer. An old and large rabbit whose meat is very tough.

Cull. To selectively euthanize or slaughter an animal early.

Process. To slaughter or kill an animal for food at the scheduled time.

Live weight. The weight of a rabbit while it's still alive.

Dress weight. The weight of a rabbit carcass that has been skinned and cleaned.

Tractor. A movable pen with an open bottom used for raising small livestock on pasture.

a barn full of rabbits. Whenever he wanted one for dinner, he simply went up into the loft and caught one. Now, this nontraditional method has potential for a whole host of problems, so I am definitely not recommending you try it, but the point is: For whatever reason, even this totally outside-the-box method worked for someone somewhere. Like I said, there is no one way to farm.

But of course, some methods are more popular than others, and that's often for good reason. When it comes to rabbits, there are two common primary frameworks: *wire raising* and *colony raising*. Within each system lies infinite opportunity for customization, but we'll begin with the basics.

Raising on Wire

The mesh-bottomed cage lies at the heart of all wire-based rabbit operations. And while there is something harsh sounding about the word *cage* to the untrained ear, try not to rush to judgment. Although I've never seen an industrial rabbit farm in person, from what I've read there are plenty of reasons to be concerned about animal welfare in many of these cage-based systems. Overcrowding, unsanitary conditions, poor air quality, and general mishandling and mistreatment most certainly take place in rabbitries that use cages. However, those problems have almost everything to do with the mismanagement of those particular rabbitries, and not the use of the cages itself. In fact, while we don't use them exclusively on our farm, there are actually several compelling reasons for raising rabbits in cages.

Raising on wire is the most common method used in commercial rabbit production. In the traditional version of this system, all breeding rabbits and their offspring are kept in a series of cages with wire bottoms, hence the term *raising on wire*. These cages serve two important functions. The first is to keep individual rabbits separate from one another, which allows you to carefully control breeding—critical for maintaining good genetics, keeping accurate records, and maximizing production. The second benefit of keeping rabbits in cages is that it reduces contact between the animals. This is a priority for many commercial producers, because it mitigates the risk of disease by limiting the three most common causes: outside contact, rabbit-to-rabbit interaction, and fecal-oral (aka "poop to mouth") transmission.

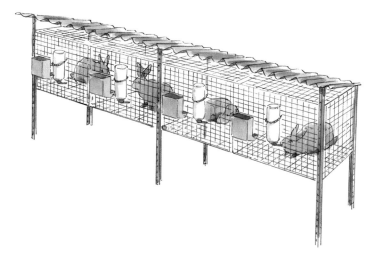

Traditional rabbit hutches can be indoors or outdoors.

The second function of the cage is to help facilitate good hygiene within the rabbitry. Wire-bottomed pens allow waste and debris to fall through the floor and onto the ground or into a collection pan, handily keeping quarters clean and dry. Damp or dirty housing can cause respiratory, intestinal, skin, and foot problems, and while it's possible to safely raise rabbits in solid-bottomed pens (with a substrate, like pine shavings or straw, to absorb urine and manure), it can be more difficult, costly, and labor-intensive to keep these types of units clean.

When it comes to raising on wire, there are all kinds of possible setups and applications. Large-scale, commercial producers generally opt for entirely metal units that are either mounted to the wall or hung from the ceiling, 4 feet (1.2 m) or so from the ground. This height gives rabbits distance from their droppings below while giving farmers comfortable access to the units for cleaning and handling. Backyard producers, homesteaders, and pet owners might use traditional outdoor wooden hutches with the same type of wire floors as the big guys. There are also stackable cages, which work best for those with limited space. In these models, cages are built on top of one another, with space below each layer for a sliding track that holds a pan. The pans, which are lined with wood shavings, cardboard, or newspaper, are designed to absorb urine and catch the manure, which gets dumped out on a regular basis.

Aside from being sanitary and convenient, raising rabbits in cages is also very efficient. Rabbits in cages grow significantly faster than rabbits raised in open pens. The reason is simple: Rabbits with less space to hop burn fewer calories. Additionally, animals raised in more climate-controlled environments expend less energy regulating themselves. And while it's pretty much true across all livestock that animals raised in confinement grow faster than those raised outdoors, we've never raised an animal in a barn or a cage for this reason alone. The benefits of growing animals more slowly, in fresh air and on grass, are just too compelling to trade. We'll take an increase in natural behaviors, improved soils, and enhanced flavor over saving a few bucks any day of the week. Something that does matter to us a great deal, however, is keeping our animals safe. Cages practically eliminate predation and contact with outside animals, reducing injury, illness, and death.

But there are downsides to raising on wire. From an economic perspective, a cage-based system can require a higher initial investment than other systems. This is because each breeding buck-and-doe pair requires its own cage, feeder, and water, and additional cages are needed for growing offspring. Wire-based systems can also require more maintenance in terms of daily chores and manure removal, especially in comparison with some pasture-based operations. But for me, the real drawback to this method is that it doesn't do much to encourage natural behaviors in the animals. Rabbits in these systems cannot dig, forage, or socialize like they would in the wild. This is a major cause of grief for many producers who, like us, strive to create as organic a landscape for their animals as possible. Generations of farmers have sought to remedy this issue, and many have settled on the second most common framework: colony raising.

Colony Raising

Colony raising is a catchall term for raising rabbits in groups or "colonies." At its core, a colony is a cage-free environment in which breeding does and their offspring live together in groups of varying sizes (generally bucks are kept separately to prevent unintended breeding). A colony may be housed outdoors in a tractor or pen or indoors in a barn,

A rabbit colony can be any open space that houses multiple rabbits, including rooms, barn stalls, and outdoor pens.

stall, or shed. The variations are endless and are often dictated by the scale of the operation and the producer's available resources.

Some of the earliest known examples of rabbit farming in colonies date back to the Middle Ages. Historical texts and maps from England illustrate domestic "rabbit warrens" throughout the countryside. These warrens were also known as coneygarths—a combination of the old English *cony*, meaning "rabbit," and *garth*, meaning "lawn." Rabbit lawns were artificially enclosed outdoor spaces cultivated for the purposes of breeding and containing rabbits for meat and fur. At the center of the traditional coneygarth is the pillow mound, a man-ufactured hill atop the earth, between 20 and 100 feet (6.1–30.4 m) long. They were made by digging a deep trench around the warren's oval- or cigar-shaped perimeter, heaping the soft dirt into the middle, and smoothing it over. Rabbits were then released onto the mounds; the surrounding trenches, now full of water, served as a moat to deter escape and prevent predation. When it came time for harvest, farmers would use handmade nets to catch the rabbits, sometimes using dogs to flush them from their burrows.[1]

Raising rabbits in this way continued throughout parts of England for a very long time, supposedly prospering until the Great Blizzard

In the Coney Garth system, rabbits are housed outdoors, in large pens with a physical barrier around the perimeter. A mobile shelter provides protection from the elements and opportunities for protected nest building.

of 1891, an event rumored to have decimated the rabbit population. Farmers did eventually recover from the blizzard, steadily rebuilding rabbit production until a widespread outbreak of *myxomatosis*, a highly contagious virus, nearly eradicated the species once more in the 1950s. Since then the industry has never fully recovered.[2]

In recent years, however, warrens have experienced a bit of a renaissance. This is, in part, thanks to the publishing of a recent experiment conducted by an American farmer funded by the Northeast Sustainable Agriculture Resource and Education program. Through conducting the research, this farmer set out to revive this traditional method of rabbit husbandry and adapt it to today's agrarian landscape. It was called the Coney Garth system, in reverence for the practice's old English roots.

In an effort to expand the potential for utility in the Coney Garth, the farmer opted to use portable barriers made of interlocking hog panels to secure the rabbits in the yard rather than the classical permanent moat. This change allowed for the rabbits to be moved on a regular basis. The farmer broke down the panels every 36 hours, moving the herd to fresh pasture. This was essentially mob grazing—a variation on rotational grazing that involves short-duration, high-intensity grazing by densely stocked animals. In this manner, the farmer was able to raise healthy, grass-fed rabbit meat and renovate scrubby pasturelands simultaneously. Constant access to new pasture also meant these rabbits could subsist exclusively on forage and therefore did not require any commercial feed.

In addition to their creative movable fencing system, the farmer designed a mobile shelter where rabbits could seek refuge from the elements (see illustration on page 36). In an effort to encourage does to kindle in a safe space, protected nesting space inside the mobile shelters was provided. Rabbits in the new Coney Garth system were able to dig, forage, and socialize without limit, hitting all the goals a wire-based system does not. It's a neat system, the details of which can be found in the project report, "The Coney Garth: Effective Management of Rabbit Breeding Does on Pasture," available for download at projects .sarc.org.[3]

While this updated Coney Garth method may be the most buzz-worthy system in colony raising today, it is by no means the only one. Countless other farmers are raising rabbits in groups, both indoors and on pasture. Not long ago I visited a rabbitry in Earlville, New York, where I met two farmers who utilize the stalls in their barn originally meant for larger livestock, like goats, to house their breeding does and bucks. In this cage-free system, groups of breeding does and their offspring live together on the ground inside the stalls, while mature bucks are kept separate to avoid unintended breeding. When they are ready to be bred, the farmers take the does to the bucks for mating and then return them to their original colony.

Because their breeding stock are kept in permanent stalls and on the floor, these folks need to be diligent about keeping the area mucked clean and dry. A single doe and her offspring have the potential to produce 1 ton of manure over the course of a year—that's a lot of poop scooping! However, since all of the rabbits are fed and watered in a single location, daily chores are very efficient. Likewise, shying away from cages and their associated accoutrements means their system requires very minimal upfront investment—a huge perk for farmers who have limited capital.

When pregnant does are ready to give birth, they are free to make nests on their own accord using any of the materials the farmers have made available. These include boxes of various shapes and sizes and a variety of substrates like hay, straw, and shavings. When the young rabbits reach six weeks of age, they are transported to an outdoor rabbit tractor, which in this case is a covered, floorless pen. Their design includes a 1-foot (30.5 cm) interior perimeter of chicken wire stapled to the edges of the tractors, which discourages rabbits from digging out. The pens are moved once or twice daily, keeping the rabbits in a clean environment and giving them access to fresh pasture.

Since they use a pelleted feed as their primary nutrition source, each tractor is outfitted with watering and feeding systems. Rabbits then supplement these feed rations with forage from the pasture. Between 12 and 16 weeks of age, the young rabbits, called *grow-outs*, are collected from the tractor and butchered. Their target live weight for this age is between 5 and 6 pounds (2.3–2.7 kg), which is 3 to 3½ pounds (1.4–1.6 kg) dressed.

Aside from low start-up costs, this second method has a few key advantages over some other systems. Raising rabbits in secure barn stalls and enclosed rabbit tractors effectively protects from predators, unlike the Coney Garth, which leaves a herd vulnerable to almost any beast of prey. Separating grow-outs from their mothers at six weeks of age and moving them to small tractors makes market-ready rabbits easy to identify and catch. Keeping rabbits in groups as opposed to individual cages allows them to socialize, while keeping grow-outs on pasture affords them opportunities for digging and foraging.

As you can probably tell, we are big fans of both colonies and pasture raising at Letterbox. We especially love how these types of systems encourage natural behaviors, increase diversity and nutrition in diet, build soil, and save money.

Challenges in Colony Systems

You might be wondering: Why doesn't everybody do it like this? That's what I wondered when I started raising rabbits, anyway. It didn't take me long, however, to learn that colony and pasture raising come with some pretty serious risks that should be taken into consideration before you dive in.

Record Keeping

There are quite a few opportunities for messy record keeping in both of the colony systems we discussed. For starters, keeping all your does together makes it challenging to know who is who. There aren't any cages to hang name tags on, so unless you've got a keen eye for subtle differences and a super-sharp memory, you'd need to tattoo or tag your breeders in order to keep track of them.

It can also be difficult to track litters in these systems. Rabbit milk is incredibly rich, containing a whopping 12 percent fat content. (For a more familiar comparison, consider the cow, whose milk clocks in at a measly 4.5 percent.)[4] This super-fatty protein-packed milk allows does to feed their kits just one or two times per day. Because rabbits are primarily crepuscular creatures (active at dawn and dusk), they usually do this very early in the morning and then again later in the evening. This means you probably won't see a doe nursing her kits, so unless you offset breeding to ensure no two does have litters in the same week, you may not be able to tell who belongs to whom. This may not be a big deal to the hobby farmer, but for the commercial producer, the effects of not being able to track your breeding can be major.

Growth Rates

As I said, rabbits that have more room to run around burn more calories. Additionally, rabbits raised exclusively on forage grow more slowly than those raised on pellets. Medium-large breed rabbits in wire-based systems typically reach a live weight of 5½ pounds (2.5 kg) in eight weeks. As I mentioned, the colony-based rabbitry I visited reached that same target weight in 12 to 16 weeks, whereas the Coney Garth rabbits in the aforementioned report needed 26 weeks, which is more than three times as long as their wire-based counterparts.

However, slow growth is not necessarily a bad thing. Studies and market surveys have shown that animals who grow more slowly and have more variety in their diets can produce meat that is both healthier and tastier. As sustainability-minded commercial farmers, we are always striving to strike that perfect balance between quality and efficiency.

Labor Requirements

While the colony in Earlville actually has very low labor requirements—aside from the regular cleaning of the stalls—the Coney Garth report documents that moving the paddocks takes one hour per day. It also mentions that it takes an additional one hour per week to round up the rabbits for regular health checks. This is significantly more time than it takes to catch a rabbit in a stall or, better yet, scoop one right from its cage. By comparison, a wire-based operation of a similar scale requires 15 to 30 minutes to execute all daily tasks associated with the operation, including feeding, watering, inspecting, and breeding. Again, this is not an argument for raising rabbits in cages—but knowing the limitations of each method empowers farmers to choose the method that is right for them (and right for the animal).

Infighting

Not all rabbits play well with others. This can be especially problematic with pregnant does and bucks, as both can become nastily territorial. Hostile rabbits can seriously injure or even kill other members of the herd in no time. However, it is possible to reduce some of this bad behavior by culling your aggressive rabbits and breeding for docility.

Higher Risk of Kit Mortality

Does and kits that are not separated from the herd are vulnerable to being attacked or accidentally injured by other rabbits. If does kindle (give birth) on pasture, as in the Coney Garth system, exposure to the elements can lead to illness and death. Even though this system provides protected spaces for the does to kindle in, many domesticated rabbits have poor instincts when it comes to nest building. Because humans have been raising rabbits in cages for so long, farmers have not prioritized these types of traits, and many of these instincts have been bred out of modern genetic lines. It is common for domesticated rabbits to build nests right on the open ground, where kits don't stand a chance against the sun, rain, wind, or cold—never mind hungry owls. While the Earlville colony reports a 10 to 20 percent average kit mortality—a pretty normal industry standard—the Coney Garth report states that of 305 kits born on pasture, only 75 lived past five weeks of age—a 75 percent loss.

Higher Risk of Illness

Rabbits raised outdoors may experience contact with wild rabbits, who can be passive carriers of *Pasteurella multocida* (a bacteria that can cause respiratory illness) among other communicable diseases. They will also be exposed to parasites that live in the soil, like Coccidia, which can lead to serious health issues over time. Disease, once present, is spread most rapidly through rabbit-to-rabbit contact and fecal-oral transmission. Colony systems encourage both of these behaviors, putting animals at greater risk.

The Wire-Pasture Hybrid Method

If you didn't before, you may now understand what I mean when I say there is no one perfect way to farm. For almost every benefit there lies a trade-off. Our goal at Letterbox has been to straddle the line between providing the most natural environment for our animals while still negating the most serious risks. After all, how happy can an animal really be on pasture if it's constantly battling illness?

We knew our rabbitry would have to strike a balance three different ways in order to be sustainable for the long term: once for our animals, again for our farmers, and lastly for our business's bottom line. Wire raising was easy on the wallet, but it didn't offer much for the animals' contentment. Colony raising was good for the animals and the farmers, but challenges in record keeping and slower growth made this method a labor of love rather than an economic opportunity. Coney Garth, while cost-effective thanks to limited infrastructure and greatly reduced feed costs, was labor-intensive and the risk for mortality was too high. It took a couple of years and a lot of trial and error, but in the end we settled on what we call the *wire-pasture hybrid system*.

In our attempt to prioritize good and minimize harm, we decided to take the best of each method and try to ditch the worst. In the wire-pasture hybrid system, we keep our breeding stock on wire only in order to mitigate degenerative pasture-based diseases (like parasitic infestation) and communicable illnesses (like viral outbreaks), which greatly reduces kit mortality. Since each doe-and-buck pair is kept in its own cage, record keeping is super simple and there is zero

The indoor rabbitry at Letterbox Farm.

instance of escape or predation on our breeding stock. To increase natural behaviors to the greatest extent possible within a wire-based setting, we invested in large cages and provide our breeders with fresh forage and lots of hay in an effort to mimic the diversified diet obtained on pasture.

Between six and eight weeks of age, grow-outs are moved outside into pasture pens, or rabbit tractors as we call them, where they stay until they are big enough to process. Exposure to *Pasteurella* is still a risk in the wire-pasture hybrid system, but parasites are much less of an issue. This is because, in our experience, mild infestations often take several months to cause health problems in healthy, young rabbits. Except in the case of a particularly bad outbreak, our rabbits have already been processed by the time something like coccidiosis would fully manifest itself. This is why it makes sense to keep our shorter-lived growing rabbits on the ground but not our breeding stock, whom we keep for two years or more.

The tractor design we use is featured in greater detail later in this chapter, but to summarize: We have several 6 × 9-foot (1.8 × 2.7 m) pens, each housing up to four litters (24 to 30 rabbits) at a time. The floor of the tractor features 2 × 4-inch (5.1 × 10.2 cm) welded wire, which prevents digging out while still allowing rabbits to forage and waste and debris to be left behind with each move. They are solid and sturdy but easily moved by one person. Some sources claim that rabbits will not eat vegetation that is lying flat due to flooring. In our experience, this is totally untrue. So long at the pasture is not too long or coarse (so no unmown hay fields), our rabbits have no problem spending the day tugging their treats through the wire.

There are endless options for how you can build your rabbit tractor, and many of them will work just fine. Just be sure that whatever you build is:

Portable. You want your tractor to be easy to transport without injuring the animals inside or the farmer doing the moving.
Covered from the elements. Your tractor should be well ventilated, provide shade, and protect the rabbits from the rain and wind.
Inescapable. There are several methods for preventing dig-outs while still providing forage. Welded wire, slatted wood floors, and chicken wire perimeters have all been used effectively at different farms I have visited.

We use 2 × 4-inch (5.1 × 10.2 cm) box wire on the bottom of our pens and move our tractors once per day, every day. The first few

Rabbit tractors in a row.

Curious grow-outs settling into their new quarters.

moves for a new group of rabbits can be a little challenging, but they quickly learn to stand on the wire and enjoy the ride to fresh snacks. However, it is important to move slowly and be observant, as you do not want to injure anyone in the process.

There is no need to sex your rabbits or separate them by gender in your tractors. If you, like us, slaughter your rabbits at 16 weeks or younger, they'll be too young to breed at this point, and accidental pregnancy will not be an issue. However, if you plan to grow your rabbits beyond 16 weeks, you will want to separate the does from the bucks to avoid unintended mating.

Unless we are experiencing an outbreak of disease, we do not sanitize our tractors between batches; we just clean out the feeders and waterers. The sunshine typically does the rest of the disinfecting for us.

It is good practice to keep running your rabbits on new land for as long as possible, but we find that we can usually get away with a 30-day rotation before returning to the first pastured area. If you are experiencing a persistent disease outbreak, sanitize your tractor once it's empty and move it to fresh ground, ideally where no rabbits have been for a year or more. This should be enough to end the disease cycle for the next batch.

When bringing together multiple litters in a tractor from their separate cages, we sometimes notice they experience a bout of running around while they feel one another out and adjust to their new surroundings. This is normal, and the frantic energy should drastically slow down after a few hours. We have not experienced much fighting within our colonies, because our entire herd has a mellow disposition. However, on rare occasions we have had a bad apple. If this happens to you, just separate the instigator into an empty cage or tractor, and resume business as usual.

Housing Rabbits in the Pasture-Wire Hybrid System

Developing our rabbit housing has been a long journey. In the beginning we used a silly-looking hodgepodge of backyard hutches gathered from forgotten corners of our neighbors' old barns. They were all different heights and sizes and built from whatever scrap materials their thrifty builders had lying around. From there we built our first rabbit tractor. It was a weird little triangular number, big enough for seven or eight rabbits, with a protected compartment for a pregnant doe to kindle in. The idea was to build one of these things for each of our does, so they could live their whole lives on pasture. It didn't work. They were awkward, they were expensive to build, and as you've probably guessed we did not have any luck getting those mamas to rear healthy litters out on the pasture. Once we landed on our current farm, we knew it was time to build something better.

Cages

Knowing that we wanted to keep our breeding stock on wire, we invested in some brand-new wire cages, purchased from KW Cages, a rabbit equipment company in California that ships all around the country. Ours are 36 inches (91.4 cm) wide, 30 inches (76.2 cm) deep, and 18 inches (45.7 cm) high. This is large enough for growing out an average-sized litter alongside the mother if need be, something

Our buck Brunswick in his home.

we do when space is tight during the winter months. Because we use these cages for kindling, we buy the Baby Saver models, which have tighter wiring on the floors to prevent leg injuries in growing kits.

These cages are all wire and arrive packed flat along with a bag of J-clips to secure the pieces together. They take just a few minutes to assemble; the only tool you need is a pair of J-clip pliers and some snips to cut out the hole for your feeder. This style cage does not come with any legs or other type of platform, however, so you'll need to decide how you want to elevate them from the ground. Some options include hanging the cages or building table frames (without tops, so the manure can still fall to the ground) to set them on. A quick online image search will demonstrate just how limitless the possibilities are. Whatever you settle on, be sure to factor the cost into your enterprise budget (more on this in chapter 12).

Today our cages hang from the purlins in our stockhouse, which is a 96 × 30-foot (29.3 × 9.1 m) greenhouse from Rimol. We use a simple system of brackets, double loop chain, and S-hooks to do this. I love it because there is nothing underneath the cage to obstruct debris from falling to the ground. Before we were able to hang our cages,

Baby Saver model cages have tighter wiring on the floor.

we simply propped them up on cinder blocks. This was an acceptable temporary solution, but poop and fur built up wherever the cage rested on the blocks, and it was a pain to clean underneath. While balancing on cinder blocks works in a pinch, I would not recommend this as a long-term setup.

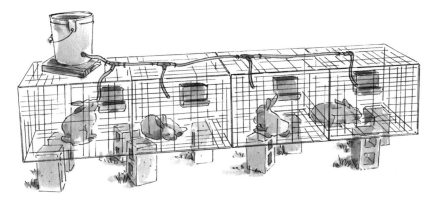

Our temporary setup worked in a pinch but wasn't ideal.

Attaching our cages to the chain with S-hooks allows us to take them down in seconds. We designed it this way so we could break down the entire rabbitry and clean out the stockhouse with the tractor if we wanted to. No tables or legs means all the debris falls right through the wire, keeping our cages clean and dry all the time. We have one cage for each breeding doe and buck. We also have a few extra for growing out litters whenever we need more space. While the number of extra cages depends on the average size of your litters, one for every four does is a good baseline.

Pasture Pens

While our first weird, triangular tractor did not work for our rabbits, it did work nicely for the quail we raised for a few years. Today, however, that tractor is officially decommissioned on our now quail-less farm, but who knows what need it might meet in the future. A lot of our livestock infrastructure works that way, starting as one thing and ending up as another, and our current rabbit tractors are the perfect example. Back when our meatbird enterprise was much smaller than it is today, we built a few of John Suscovich's Stress-Free Chicken Tractors to house

Once you pop on the wheels, these tractors are super easy to move.

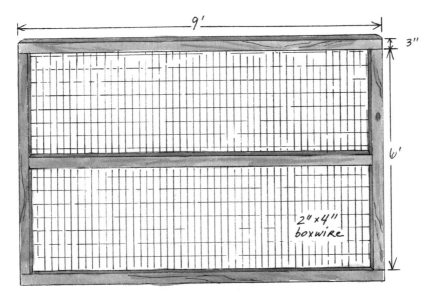

A simple alteration transforms an open-floor chicken tractor into an escape-proof rabbit pen.

our broilers. It's a neat little design intended to hold 30 or so chickens, and it worked well for us for a couple of seasons. As we began to scale our meatbird enterprise, the design proved too small for our needs, so we retired the model and built something more appropriate. As it turns out, John's Stress-Free Chicken Tractors are the perfect size for growing rabbits, and with a few minor adjustments we were able to convert our old chicken tractors into our new rabbit pens.

We love this style pen for a couple of reasons. For starters, it's strong and sturdy but still easily moved by one person. The high-arched design allows for maximum ventilation while providing the rabbits with shade and protection from the elements. They're affordable and, if built correctly, will last for many years. Ours are more than five years old, and they still look brand new. All we have needed to do over the years to maintain them is replace a tarp here and there. The other reason we love these tractors is because they're nice to look at, which, while not paramount on our farm, is always our preference.

You can find the plans for this style tractor in John's e-book, available on his website.[5] It's well worth the $10. To convert it to a

Rear view of the rabbit tractor.

rabbit tractor, we simply ran a 2×4 piece of lumber down the center, longways, and stapled on 2 × 4-inch (5.1 × 10.2 cm) box wire to keep the rabbits from digging out. The box wire conveniently comes in 3-foot-wide (91.4 cm) lengths, which is perfect for our 6-foot-wide (1.8 m) tractors once they're divided down the center. This style tractor is 6 feet wide by 9 feet (2.7 m) long, which I find houses 24 to 30 rabbits, or four average litters, quite nicely.

Feeders and Waterers

Our plans include one 68-ounce (1.9 kg) Siftomatic metal feeder for each cage and two mounted inside each tractor, also from KW Cages. For water, we use 5-gallon (20 L) buckets attached to 5/16-inch (7.9 mm) plastic tubing with a nipple water attachment on the end. The nipple water attachments are very inexpensive and can easily be found online. I always keep a bag of extras on hand, as they tend to gum up or leak and need to be replaced. We check each one, every day, to make sure they're all working properly.

We use one bucket per 10 cages in our barn and one bucket per tractor in the fields. In the tractors, we use plastic tees to split the water from the bucket to two different lines, so multiple rabbits can drink at a time.

We also have a pen cup per cage, which we fill in the winter when our water lines are prone to freezing and in the summer as a backup water source for when it's really warm. Don't buy these at the pet store—they're really expensive there. Instead, purchase from a live-stock supplier, like KW, where they are about a dollar each. I like the tough plastic ones that clip onto the wire.

So there you have it, the system that works best for us! I hope it will work for you as well. Fear not if you choose another method instead or decide to invent your own, though. Most rabbit husbandry basics remain the same across all systems, and in the following chapters we will walk through breed selection, feed and water requirements, breeding, housing, processing, and more. Each one of these husbandry basics can be tweaked to fit any system discussed here and likely anything new you might dream up yourself as well.

CHAPTER 4

Choosing Your Breed

I have been obsessed with animal breeds since way before I ever considered a career in livestock farming. If my childhood memories can be trusted, my interest in this particular science was first born while watching the original *101 Dalmatians* when I was little. In the movie there is a scene where Pongo, the canine lead, is looking out the window of his owner Roger's apartment. Down below, humans who resemble their pets parade a variety of iconic, purebred dogs down the road. There's an Afghan hound striding alongside a tall, lean woman with long, straight hair, an elegant poodle leading a haute Londoner, and a little, round pug trotting beside a little, round lady. The fact that somehow, over the span of a millennium, humans managed to take some wild canid and, from it, magically develop the perfect companion for each one of us, wildly different as we are, had me awestruck. Today I still think it's all really cool, and I love learning about what makes each breed of animal special.

The technical term for breeding and raising domestic rabbits is *cuniculture* (*cuniculus* is the Latin word for "rabbit"). As far as we can tell, humans have been practicing cuniculture for at least 1,500 years, and over the course of that millennium and a half we have managed to create more than 300 breeds of rabbits throughout 70 countries. Of these, the American Rabbit Breeders Association (ARBA) recognizes

Angoras are a fiber breed of rabbit, known for their soft wool. *Photo by gygyt0jas.*

49, which, for the cuniculturalist's purposes, can be divided into three categories based on usage.[1]

Rabbits for Wool

Most people think of sheep when they think of wool, but in actuality this soft, curly, ever-growing hair can come from many different animals. Goats, alpacas, cattle, and even camels are used to manufacture wool all around the world. When it comes to rabbits, the Angora is the most commonly used breed for wool production. Producers harvest their long hair by shearing it or combing it loose. The wool from Angoras is exceptionally fine and renowned for its softness, warmth, and fluffiness. Jersey Woolies, Lionheads, and American Fuzzy Lops are also wool-breed rabbits, but because of their small size, they are typically kept as pets and show animals rather than for fiber production.

Because these types of rabbits have been bread for fur quality rather than meat quality, I would not consider these breeds if meat

Flemish Giants are the largest breed of domestic rabbit. *Photo by Vronja_Photon.*

production is the primary goal. Wool rabbits require regular brushing and other special treatment that meat breeds do not, so while meat could very well be the by-product of a wool operation, wool is unlikely to make sense as the by-product of a meat operation.

Rabbits for Fur and Pelts

Some rabbits have been specifically bred for the purpose of fur trading. Fur rabbits typically have short, shiny coats rather than the long, fluffy hair of their wool-breed counterparts. Their pelts, which are skins with the fur still attached, can be used for hats, ballet shoes, gloves, and many other clothing items. Rex, Satin, and Silver Fox rabbits are the most commonly used breeds for fur production because of their large size and desirable coat qualities. Fur rabbits must be raised to maturity (at least five months) in order to obtain pelts that are large enough and of high enough quality. Both rabbit wool and rabbit fur markets are marginal in the United States, so most producers of these breeds are hobbyists.

Rabbits for Meat

The primary characteristics of a good meat rabbit breed are threefold. First, they must grow quickly and efficiently. Second, they need good mothering skills so they can routinely raise litters of eight or more kits. Last, they have to grow to the right size, with a good meat-to-bone ratio. In the United States meat rabbits are typically raised to be 3 to 4 pounds (1.4–1.8 kg) dressed, with small bones. In order to be considered large enough for meat production, mature rabbits should weigh between 9 and 12 pounds (4.1–5.4 kg). New Zealand White and Californian rabbits are the most commonly used breeds; however, 14 others are also considered viable for meat production. Let's take a look at nine of the most commonly used.

American Rabbit

The American rabbit was first recognized by the ARBA in 1918 when it was still known as the German Blue Vienna. Today it is considered a *dual-purpose* breed, which means that it can be effectively used for both meat and fur production. The fur of an American can be blue or white, with the blue variety being the deepest blue color of any of the breeds in the United States. Despite its rampant popularity throughout the first half of the 20th century, the American rabbit is now one of the rarest breeds in the country. However, at their best, these rabbits are large, docile, and fast growing, with good mothering instincts, which makes this heritage breed ripe for a revival.[2]

American Chinchilla

The American Chinchilla first gained recognition in the early 20th century for its striking resemblance to the South American chinchilla, an adorable little rodent from which this rabbit got its name. They are considered large-breed animals, with adults weighing 9 to 12 pounds (4.1–5.4 kg), and are known for having good meat-to-bone ratio. Although today their status is critical, the once wildly popular American Chinchilla holds the record for most ARBA breed registrations in a single year. Likewise, it has contributed to the development of more breeds worldwide than any other rabbit.[3]

Giant Chinchilla

The Giant Chinchilla originated in the United States when cunicul-turalist Edward Stahl crossed Chinchilla rabbits with Flemish Giants. It has since been purebred for over 45 years. The Giant Chinchilla is an extra-large rabbit, weighing up to 16 pounds (7.3 kg), with a docile nature. They are a popular rabbit for backyard meat producers because they grow very quickly and can reach 7 pounds (3.2 kg) in as little as eight weeks. However, due to their heavy stature, Giant Chinchillas can be prone to developing sore hocks when raised on wire, making them less desirable for commercial production. The Giant Chinchilla is sometimes called the Million Dollar Rabbit, because Edward Stahl actually became a millionaire from selling their breeding stock.[4]

Californian

The Californian, or Californian White, rabbit was developed in California in the early 1920s by a breeder named George West. He began by crossing purebred New Zealand Whites with Chinchilla and Himalayan rabbits, the latter being the source of their distinctive markings. Today these are by far one of the most popular breeds for commercial production, and

Though a mixed-breed, this young rabbit looks most like its Californian mother.

they are used for meat and pelts as well as household pets. They are a hearty breed known for being fast growers and great mothers.[5]

Champagne d'Argent

The Champagne d'Argent rabbit is one of the oldest pure breeds. While the exact origin is unknown, it is believed to have originated in the Champagne region of France at the beginning of the 17th century. These big bunnies have lustrous coats, which makes sense as *argent* is French for "silver." While not terribly popular in the United States, these rabbits are common worldwide and are very well suited for meat production.[6]

Flemish Giant

Flemish Giants are extra-large rabbits that can grow to be a whopping 20 pounds (9.1 kg) at maturity. According to the *Guinness Book of World Records*, the longest rabbit ever recorded is a Flemish Giant named Darius who clocked in at 4 feet 3 inches (129.5 cm). These gentle creatures are very popular pets, but despite their size they are not common among meat producers. This is because they are slow to reach maturity, and while young Giants do actually grow quickly, they mostly put on bone during their first 70 days rather than meat. By the time the meat-to-bone ratio improves, they are typically too old to be considered *fryers* (the market term for young and tender rabbits) and are therefore harder to market in the United States.[7]

New Zealand White

New Zealand Whites are the second most common breed used in commercial production, behind the Californian. Despite their misleading name, they also originated in California in the early 20th century. They have a genetic deviation known as albinism, or a lack of melanin (which gives the skin, fur, and eyes their color). This is why New Zealands have their trademark red eyes, pink noses, and snow-white pelts. This breed is generally healthy, hearty, and quick growing, with good fur and meat quality. Full grown, they are 9 to 12 pounds (4.1–5.4 kg).[8]

Satin

Satin rabbits are known for their silky, lustrous coats that come in a wide range of colors. They originated in Indiana in the 1930s and are

Another mixed kit displays its Satin ancestry.

derived from Havana rabbits, which had a genetic mutation that caused a hollow hair shaft. This hollow hair shaft is what gives their coats the unique shine that inspired their name. They are productive medium-to-large rabbits that have a history of good breeding that contributes to their good health today. Satins are considered medium-large rabbits and weigh 8 to 10 pounds (3.6–4.5 kg).[9]

Silver Fox

The Silver Fox is the third oldest breed of rabbit developed in the United States. Despite a recent uptick in their popularity, this breed is still considered threatened by the Livestock Conservancy. Their silvery-blue fur is unique in that, when stroked from tail to head, it will stand straight up until it is stroked back in the other direction. This characteristic does not exist in any other breed of rabbit but is also found in the pelt of the canine silver fox of the Arctic (where the breed name originated). Silver Fox rabbits are uniquely American and do not exist outside of the country.[10]

———

On our farm, our does are primarily Californians but we also have a few New Zealands and some Satins. For bucks we have one who is half Californian, half New Zealand, and another purebred Champagne d'Argent. This means that all of our juvenile rabbits are "crossbreeds," rather than purebreds. This method of crossbreeding has its own strengths and weaknesses, which I discuss in chapter 6.

CHAPTER 5

Handling Your Rabbits

Ll of my agricultural mentors have been primarily vegetable farmers, and I do not have a background in formal livestock training. I did, however, spend a couple of months volunteering on the Noah's ark of hobby farms in my early 20s. It was a little postage stamp of land in Northern California that had two of everything—more like a petting zoo than anything else. The whole project was casual, to say the least, and I probably picked up more bad habits than good ones during my brief tenure there. However, having the opportunity to be around so many kinds of livestock did help me start building one important skill: the ability to move around animals.

At this point in my farming career, handling livestock is a skill that is so innate that most of the time I forget it's a skill at all. What reminds me this is a valuable and teachable technique is watching a non-farmer or a new crew member try to catch a chicken for the first time. They'll chase it around and around forever, paying no attention to how it's moving. Don't get me wrong—I used to chase chickens around for hours, too, and even used to be terrified of pigs. I remember struggling forever with rakes, trying to flip overturned feeders right-side up from outside the pigpens because I was scared they would bite me if I stepped inside (what I used to mistake for swine-y aggression, I know now is just unbridled joy and buoyant curiosity). Today I can catch a chicken in seconds because I know where it's going before it does.

When it comes to handling your rabbits, your instincts will develop naturally as you spend time with them. The more you handle them, the more amenable to handling they will become. There are, however, some basic rules to follow.

Handle Your Rabbits Thoughtfully

Rabbits are prey animals, so they startle easily. Scared animals are stressed animals, and stressed animals become sick animals in no time. You don't have to move super slowly, but no livestock I know responds well to erratic or sudden movements. Keep your motions fluid and your actions gentle but firm.

To pick up a rabbit, grab hold using the skin on the back of the neck, known as the *scruff*. You can hold a full-grown rabbit by the scruff for a few seconds, but if you need more time than that, you will need to

Handling a medium-sized rabbit.

Holding a small rabbit by the haunches. *Photo by Nichki Carangelo.*

support the weight of the rabbit with your other hand. You can support smaller rabbits by the tush, but larger rabbits require more stability. I usually saddle full-grown rabbits onto my forearm, supporting their hindquarters with the palm of my hand. I tuck their head gently into the bend in my elbow. In most cases, our rabbits will stay calm in this position as long as I need them to be there, but once in a while one will start to struggle, scratching the heck out of my arms. When this happens, I usually grab the rabbit by the scruff and hold it out away from my body for a second or two. When it calms down, I reposition it back onto my forearm. This usually does the trick. If I lose control of a rabbit, I find it's best to just let go, allow it to drop to the ground on all fours, and then pick it up again. Struggling too hard to hold on to a finicky rabbit is stressful and can cause injury to you, the rabbit, or both. Always restrain a rabbit using the least amount of force possible, and despite what you've seen in cartoons, don't ever pick up them up by their ears. Young, light rabbits can be held upside down by their haunches without issue, though.

Along with the various times you will be moving them around your rabbitry and farm, you will need to handle your rabbits on some other

Handling a rabbit for an extended period of time.

WELLNESS CHECK

On our farm, we typically check for 10 things when performing a regular wellness check:

1. **Alertness.** This is the first thing I check all of our animals for every single day. Healthy animals are interested in their surroundings, even when they're calm and contented. While rabbits are allowed to get sleepy, take notice if they're overly lethargic or non-responsive to environmental stimuli. A lack of alertness is an indication that there is an issue.

2. **Overall body condition.** Healthy rabbits aren't too fat or too thin. To check for good conditioning, start by running your hands over the back of the rabbit. You're feeling for the hip bones, ribs, and spine. Just as in dogs and cats, these things should be easy to locate but feel rounded, not be bony or acute. If it's difficult or impossible to feel the ribs, your rabbit is overweight. If the ribs feel sharp, like rulers, your rabbit is underweight or, in extreme circumstances, emaciated.

   Many rabbit producers, particularly those raising for show, use a specific 1 through 5 scoring system when checking for body condition. In these systems, a rabbit that scores a 1 is emaciated, a 3 is ideal, and a 5 is obese. If you're new to raising rabbits, it can be helpful to consult these guidelines while honing your own judgment. Look for resources from 4-H or rabbit show clubs.

   You can also check to see if your rabbit is properly hydrated with a *tent test*. To perform, take the skin on the back of the animal's neck and stretch it upward until it's just taut and let it go. It should snap back to its original position immediately. If the skin is slow to snap back, your rabbit is likely not getting enough water.

3. **Eyes.** At this point in my career, I can identify an unwell animal in seconds, just by looking at their eyes. Healthy animals have clear, bright eyes, and while this looks different between breeds and species, you'll be able to recognize when something's off after

you gain a little experience. In the meantime, check to make sure they're clear of dullness, redness, cloudiness, or discharge.

4. **Nose.** Watch their nose to make sure it's twitching regularly and not runny. It should be clean and dry. Rabbits don't sneeze very often, so a sneezy bunny should be paid extra attention.

5. **Paws.** Check the foot pads for cracks, sores, or lumps. In furry-footed rabbits, make sure they're clean and free of any matting. Your rabbits' nails shouldn't be too long or cracked. If they are, it's time to give a quick trim.

6. **Ears.** Check for scabs or waxy buildup as far as you can see into the ear canal. The presence of either can be an indicator of ear mites. If you notice that the tips of a doe's ears have been chewed on by her babies, it probably means it's time to move the little guys to another cage or out to your movable shelter.

7. **Teeth.** Check to make sure they aren't overgrown and that the top and bottom teeth are wearing evenly. The top and bottom teeth should meet but not overlap.

8. **Fur and skin.** Their coat should be shiny and smooth. Check for matting, patchiness, scabbing, scratches, or lumps by quickly running your hand over each part of the body. If your rabbit is molting or just has a lot of loose fur, this is a great time to give it a quick brushing. Keep an eye out for excessive dandruff—it can be an indicator of fur mites.

9. **Vent and scent glands.** Check to make sure the *vent* (the general area that includes the sex organs and anus) isn't swollen, inflamed, or scabby. There is a scent gland on each side of the vent. These two glands secrete a waxy substance that can build up and become impacted if the rabbit does not clean itself properly. If you notice a buildup, wipe it away using a Q-tip dipped in Vaseline.

10. **Rear end.** Check for dirt or fecal matter on or around the underside. Rabbits are fastidious self-groomers, so the presence of dirt or poop can be an indication of illness or dietary issues.

If everything on this list looks good, you have a healthy rabbit on your hands.

occasions—for example, performing regular health inspections on your breeding stock. This is a good idea for two reasons: It gets your rabbits accustomed to being handled, and it allows you to catch any small issues before they become major problems. We keep a small, counter-height table in our stockhouse that we can use for our rabbit inspections. Make sure the surface of your table is slip-free. Rabbits, even when they are being held still, feel unsafe on slippery surfaces. You can put them on top of a towel or a couple of flakes of hay. In a pinch, I have used my sweatshirt, and it's worked just fine. You should check all of your rabbits thoroughly at least once per quarter.

Nail Maintenance

To trim your rabbits' nails, you can use special dog or small animal trimmers or a regular pair of household nail trimmers. Be careful not to cut the *quick*, which is the portion of the nail containing blood. It can be hard to see in very overgrown nails, but you can use a penlight or your cell phone flashlight and shine it through the nails—the quick is the area that looks pink or cloudy. Mark it with the thumb and index finger on your weak hand and clip just above it with your other hand. If you end up nicking the quick and it start to bleed, don't panic, even if it seems like a lot of blood. Just apply pressure with a clean towel until it stops bleeding. You can also use a styptic powder like Kwik Stop to stop the bleeding, but I know a dog breeder who swears it's better to dip the paw into a cup of clean dirt instead. I can vouch for the dirt method—it works!

Sexing Your Rabbits

If you know what to look for, you can identify the sex of your rabbits as early as three weeks of age. However, it is much easier to do so when they are mature. As this stage, they will begin to show some characteristics of what will eventually be fairly pronounced *dimorphism*, meaning the males and females have clearly visible, different physical characteristics. Bucks have bigger, blockier heads than does

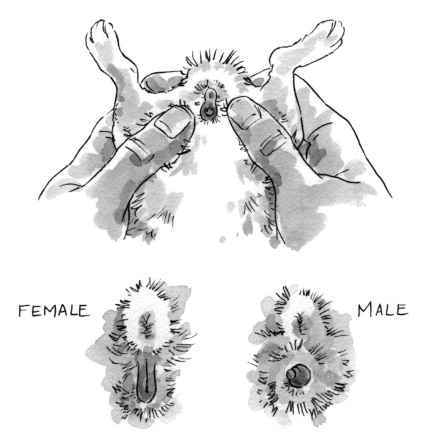

FEMALE   MALE

Sexing a rabbit is easy, once you know what to look for.

(like bulls compared with cows), and their bodies tend to be squarer. Mature does are usually larger than their male counterparts, but the real giveaway for an adult female is her *dewlap*, the adorable floppy extra skin below her neck.

However, to definitively sex a rabbit, you will need to check its genitals. To do so, start by flipping the animal on its back, either by cradling it in your arms or by laying it down on a table. When you look at the area below the tail known as the vent, you will see two openings. The one closer to the tail is the anus, which is where waste is eliminated. The other opening is where the genitals are. Press down on both sides of the vent using either your thumb and forefinger or forefinger and middle finger. The genitals will protrude, revealing a pinky-red, moist organ, which is either the penis or vagina. To determine which it is, look carefully for either a slit or a circle within the protrusion. If you see a slit, the rabbit is female. If you notice a circle, it's a male.

Transporting Your Rabbits

We use poultry crates anytime we need to move our rabbits any distance. They are easy to load and super secure, stackable, and durable. They also have open bottoms, so poop and pee fall through, eliminating the need for bedding. We can usually fit a dozen or so

We use poultry crates to transport our rabbits.

16-week-old rabbits per standard poultry crate at a time. If you don't want to invest in this type of crate (the good ones are about $80 each, including shipping), you can use dog and cat carriers lined with straw, hay, or shavings to transport your rabbits.

Catching Your Rabbits

It's easy to catch rabbits when they are in a cage, but it's a bit more challenging in pasture-based systems. When it's time to load our rabbits from out in the field, we start by moving the tractor to fresh grass. We do this for two reasons. The first is so that when someone is inside the tractor, they're not stepping on any poop that has accumulated since the previous move. While we don't practice the most fastidious biosecurity at Letterbox, our team does our best to avoid tracking poop and all of its potential for accompanying parasites from one place to another. The second reason we move the tractor first is because new grass keeps the rabbits distracted for at least a few minutes.

Most of the time, at least a few curious rabbits will gather at the door as soon as we open it, and they are easy enough to grab. To get the others, you will have to slowly enter the tractor. Be sure to close the door behind you, unless you want to spend the afternoon chasing loose bunnies with a net. Once you're in there, crouch down, avoid sudden movements, and grab the rabbits one by one with a quick grab of the scruff. So long as you move swiftly and deliberately and avoid unnecessary flailing or stomping around, you should have no problem scooping up however many bunnies you need to. If for whatever reason you startle the rabbits and they start running in circles or frantically climbing the wire, calmly exit the tractor and wait for them to settle down before entering again. This should only take a few minutes, and trust me, it's worth the wait. Trying to catch rabbits when they are worked up is almost impossible, and frenzied rabbits can easily injure themselves.

As embarrassing as it is to admit, I have been known to let some livestock loose on the rare occasion. Sometimes it's because a tree limb falls on the electric fence we use for the pig run, but 90 percent of the time it's because I get cocky and decide not to close the door behind me while I'm loading rabbits or chickens. Another 5 percent of the time it's because I forget to double-check the latch after I do close the door. If your rabbits escape from the tractor, my advice is to act fast. For the first 10 minutes or so, the rabbits are wary of their newfound freedom. They explore the novel world around them very cautiously at first, wanting to be sure nothing will kill them before taking each slow hop. This brief window of bunny insecurity is your best opportunity to catch them—do not waste it. Once your rabbits find their bearings, they'll be a heck of a lot harder to catch. If available, I recommend grabbing a long-handled net and a couple of friends to assist in the wrangling process. With enough effort, you will get them eventually. If there are any extra-clever escapees still on the lam, you can set a Havahart trap right up against the outside of the tractor. Odds are your fugitives will return to scene of the crime soon enough.

If you live somewhere with a robust wild rabbit population like we do, and your loose rabbits run free for more than a few hours, I would not put any of them back into the tractor with the rest of your herd once you do catch them until you're certain they are still healthy. There is a chance they found some wild rabbits during their time in the wild and may have picked up one of the illnesses they carry. To be safe you may want to quarantine any rabbits that have been loose for an extended period of time, to make sure they are still healthy.

CHAPTER 6

Breeding Basics

As I mentioned, we raise a combination of purebred New Zealand Whites, Californian × New Zealand crosses, and New Zealand × Satins. The *x* in the latter two stands for "cross" or "crossbreed." A *crossbreed* is an animal that has purebred parents of two different breeds from the same species. So mating a purebred New Zealand with a purebred Californian will result in New Zealand × Californian offspring. A *mixed-breed*, on the other hand, is what we call an animal with unknown parentage or parentage from other crossed or mixed animals. Neither of these terms should be confused with *hybrid*, which is used to denote the offspring of two different species—like a mule, which is the result of mating a donkey and a horse.

Who to Breed

If you are planning on raising rabbits for show or are working to preserve a rare or endangered breed, you will likely want your animals to be purebred. When raising rabbits for meat, however, this isn't so important. Both pure-breeding and cross- or mix-breeding have their benefits and drawbacks. When bred correctly, purebred animals will reliably demonstrate the same desirable characteristics generation after generation. This is why so many people want a pure-bred dog—they know what they are going to get. My clownish, loyal, boundary-conscious Old English Sheepdog is exactly the herding dog I expected her to be. But crossing breeds can result in something called *hybrid vigor*. Hybrid vigor, or *heterosis*, is the tendency of a crossbred animal to demonstrate qualities superior to both parents. In rabbits these qualities can include things like general health, growth rates, and meat-to-bone ratio.

As I mentioned in chapter 4, we practice crossbreeding on our farm. This is just because I feel more comfortable hedging my bets on the natural occurrence of heterosis than I do trusting my own ability to pick out rabbits with the very best characteristics. If I had more formal training in judging rabbits, though, I might very well raise purebreds instead.

When to Breed

Knowing when your rabbits are ready to breed will depend on the type of rabbit each is. Because their ability to reproduce is contingent on size rather than age, it is different for each one. In technical terms, bucks are not actually considered sexually mature until the moment when their daily sperm production stops increasing. For large breeds like ours, this is around 32 weeks. However, they can start exhibiting mating behaviors as early as eight weeks. If you see a young buck mounting a doe at this stage, though, don't worry—at this point he is probably still physically incapable of impregnating her. It's closer to 20 weeks of age that, though still a couple of months away from technical maturity, bucks become capable of siring viable pregnancies.

Does, on the other hand, typically reach sexual maturity upon reaching 70 to 75 percent of their adult body weight. For larger breeds like ours, freely fed does will reach this prerequisite 7 or 8 pounds (3.2–3.6 kg) sometime between 20 and 36 weeks of age. Because breeding small does can stunt their overall growth, many producers recommend waiting until they have reached 80 percent or more of their adult weight before breeding. While does younger than 20 weeks might appear to accept a frisky buck, it is very unlikely they will be ovulating yet and thus cannot get pregnant. In general, the smaller the breed of rabbit, the younger they become fertile. Conversely, the larger the breed, the later they become fertile.

The Breeding Process

Let's assume your rabbits are old enough, and you are ready to make some more. If everyone is in good health and feeling cooperative, breeding rabbits is very easy. This is because rabbits, uniquely, do not have an *estrus*, or heat cycle. Mammals with estrus cycles are only sexually active while they are ovulating and not during the period of time in between. The length of time between estrus cycles depends on the type of animal. In sheep, for example, estrus occurs seasonally every 17 days, while for pigs it's every 21 days. This means sheep and swine breeders, among others, need to take notice and track their females' heat cycles and place them with a male during a specific time. Luckily for us, breeding rabbits is much simpler. Does release eggs at the onset of intercourse, rather than in sync with particular hormone cycles. This means they can, in theory, be bred at any time. In fact, a doe is considered to be in heat anytime she accepts a buck and is considered to be in *diestrus*, or out of heat, whenever she declines. While certain environmental factors can play a role in determining periods of estrus and diestrus, why or when a doe will refuse a buck is, annoyingly, random as far as I can tell. Fortunately, most of the time our does are receptive to our bucks.

Now that you know you can breed your rabbits any day of the year, let's get to the nitty gritty. To breed your rabbits, first take your doe to your buck. This is important. If you bring the buck to the doe's cage instead, he might spend too much time smelling around this new environment; the doe can become territorial instead of submissive, as you want. Once you bring the doe to the buck, she will sit still if she goes into heat, with her back arched downward and her hindquarters raised up, ready to accept the buck. The scientific name for this pose is *lordosis*, but we usually call it *lifting* because the doe's tail "lifts" up as a result of entering this stage. The buck will then mount her from behind and do his thing. The act of intercourse takes less than

The buck mounting the doe.

30 seconds, and a successful mating will usually end with the buck ceasing movement and falling off the doe onto his side. If you've never seen what a successful breeding looks like in real life, just take a look on YouTube (how wonderful it is to farm in the 21st century, sometimes). If the doe does not lift for the buck, and instead hunkers down in the corner or runs around and around, she is not in heat, and it is unlikely that she will successfully breed at this time.

How long you leave the doe with the buck is up to you. Some farmers remove the doe from the buck's cage after witnessing a single good take. Others leave the pair together for half an hour or more. At Letterbox we like to see two good mounts take place within a minute or two, since additional breedings can yield larger litter sizes. This has to do with how rabbit sperm works. The first time a buck ejaculates, there is a lot of semen, but it's not very concentrated. The second time around there is less but what is there is much denser with sperm. The more sperm fight their way to the eggs, the more babies. After a buck's second ejaculation, however, sperm count decreases and fertilization becomes less likely. This is why it's best to service only one doe with any given buck per day if you can.

The buck, about to fall off.

We separate our buck and doe right after two successful mounts take place. In my experience leaving them together for too long can cause them to fight, and an angry doe can injure a buck in no time at all. Given that the semen gets weaker after each take, I don't see the point in leaving them together for prolonged periods of time. Additionally, it's best to supervise your breedings for their entire duration so you can both be sure it was successful and keep an eye on everyone's behavior.

If your buck has mounted your doe, given a few good thrusts, ceased movement, and fallen off, chances are that your doe is pregnant. If he mounts her then dismounts, or she moves out from under him, it is unlikely your doe has been successfully bred (but not impossible!). Sometimes the does aren't cooperative, and they'll run around the cage while the buck follows or loses interest. This frustrating behavior is common with first-time mothers and young bucks. It can also be triggered by the season. A buck will often bite the back of a doe's neck either in an effort to persuade her to lift her tail, or—if things are going well for him—in order to keep his balance while he thrusts. Don't be alarmed by this behavior—it's totally normal.

Breeding Woes at Letterbox Farm

Like I said, breeding *should* be easy, but the truth is, sometimes it's a real pain. For example, one December we found it nearly impossible to breed almost any of our rabbits. We had just increased our herd and had a bunch of young does who had never been bred before. I was spending hours in the stockhouse moving does in with our bucks over and over again, only to watch them run in circles until both rabbits totally lost interest. Sometimes they wouldn't even run around—they would just sit there. I even tried moving the pair to a high tabletop where I would hold the doe still while the buck mounted. While a few of the mounts accomplished this way appeared successful, they never resulted in an actual pregnancy.

After consulting with some other rabbit producers, I came up with a three-phase plan of attack. I bought a new, young but *proven* buck (meaning having demonstrated successful breeding at least once), just in case the problem was with Frank, our four-year-old New Zealand buck. I added apple cider vinegar to the water, having heard that the small amount of alcohol might encourage our does to loosen up a little, and began supplementing their feed with black oil sunflower seeds. The new buck was wonderful, and he gave his best effort, but still our does wouldn't lift.

While always a good supplement, the apple cider vinegar had no noticeable effects. Black oil sunflower seeds are nice and fatty, so even though I can't say they helped with the low libido of our herd, they did do wonders for body conditioning. We now feed a tight fistful to each of our breeders most days, just because.

What did ultimately serve as a marker of improvement was the coming and going of the winter solstice. Beginning on December 22 of that year, our success rates began to climb, and by the second week of January, all of our does were breeding as usual. They have continued to breed on schedule ever since. Should you encounter this problem during the darkest days of winter, my advice to you is to turn on some regular overhead lights and ensure that your herd is experiencing 14 to 16 hours of light. And the black oil sunflower seeds certainly don't hurt.

Breeding Troubleshooting

Without artificial lights, rabbits in northern climates tend to enter a natural rest season during the late fall. If you are experiencing difficulty breeding from September to December, the short days are the likely culprit. But as I mentioned earlier, does can be in diestrus for mysterious reasons at random times. If your rabbits aren't breeding, there are a few things you can try. First, go ahead and remove the doe and try again another time. If you're lucky, she will be in a better mood in a few hours. If not, wait four or five days before you give it another go.

You can also try breeding the doe to a different buck. We typically keep two bucks at a time, and sometimes when I place a doe with one, she's just not into it, even if the buck is raring to go. But when I move the doe to the second buck, she lifts right away, and we get a successful breeding. I have no idea why a doe might reject one buck and accept another seconds later, but it definitely happens.

If you normally breed your rabbits later in the day but things suddenly aren't going well, breeding first thing in the morning might improve your odds. This is especially true on warm days, as rabbits greatly reduce their activity to keep from overheating. If it's hot, they are more inclined to stretch out and relax rather than to get busy. Studies have also shown that warm temperatures reduce the motility and potency of a buck's sperm. These effects on sperm have been observed to take place in as little as eight hours in 95°F (35°C) weather as well as during prolonged periods (14 days) of temperatures as mild as 85°F (29°C).[1] Breeding early in the morning while it's still cool will help ensure frisky bucks and active, healthy sperm.

You can also take a look at the body composition of your bunnies. If they are looking thin or bony, they may not be getting enough nutrition. Increase their feed, break out those sunflower seeds, and keep an eye on their weight. Conversely, make sure your rabbits are not too fat. Overweight bucks tend to lack stamina and can lose interest in a doe who does not lift after one or two of his advances. If your buck seems a little lazy and is looking chunky, it's time to put him on a diet. Use the target weights for your specific breed as your guide.

Overweight does may readily lift for their buck, but all that extra fat can choke off an otherwise successful mating. If your doe appears to be mating with the buck but fails to produce a litter or regularly produces very small litters (two or three kits), she could be on the heavy side.

If your rabbits aren't too skinny or too fat and nothing else is working, I have only one last piece of advice. Take a second look and make sure they're not both bucks. Hey, it's happened to us.

Choosing Which Rabbits to Mate

Now that you know how to breed your rabbits, it's time to answer the question of whom to breed them with. In cuniculture there are two common methods: *line breeding* and *outcrossing*. Line breeding is the mating of animals that are closely related in order to select for desirable attributes—for example, breeding a doe back to its sire. Outcrossing is the breeding of animals that are not closely related.

Many rabbit producers firmly believe in the benefits of line breeding, and understandably so. Educated farmers with a firm grasp on the genetics of their own herd can rapidly breed out undesired characteristics and consistently obtain desirable ones in just a few generations. In fact, every purebred rabbit is at least distantly related to every other rabbit of its same breed. Line breeding, after all, is how new types of rabbits are made.

For less experienced farmers, line breeding can end up causing more harm than it does good. In the same way professional breeders use this method to maximize good traits, a novice might accidentally lock bad traits into their herd. Once this mistake has been made, there is no way to undo it aside from bringing in new rabbits and starting over.

We generally practice outcrossing, breeding animals that aren't closely related. In our case, we rely on careful selection and hybrid vigor to produce big, healthy rabbits. Currently, we are using New Zealand bucks from two different lines and Californian does. For variety's sake, we have saved some of the best Satin × New Zealand does that were sired by our retired Satin buck and our New Zealand does.

Don't be afraid to cull your breeders if you find you have made the wrong selection. As was mentioned in the beginning of this guide, uniform and reliable production is imperative to running a successful rabbitry, and nothing affects this more than developing quality breeding stock. Since we are breeding for commercial production and not for show at our farm, our baseline for performance is fairly simple. The three criteria for our breeding rabbits are as follows:

1. They must be easy to handle and breed. This means consistently lifting or mounting within 30 seconds of being introduced.
2. They must be good mothers who produce a minimum of six kits per litter and ideally eight or more.
3. They must be part of a genetic line that is healthy, does well on pasture, and finishes between 2¾ and 3½ pounds (1.3–1.6 kg), dressed, in 12 to 16 weeks.

We do not breed for pelts on our farm, but if you have a market for them in your area, you may want to consider breeding for color and fur quality as well.

Techniques for Detecting Pregnancy

Being able to accurately detect a pregnancy (or lack thereof) will save you time by allowing you to rebreed any doe who's not pregnant without having to wait 30 days to be sure. Since we all know that time is money, here are a few quick tips.

First, immediately after mating, flip over the doe and check to see if her vent, or genital region, is glossy. If it is, she is most likely bred. If not, rebreed her within the next 24 hours.

Between 10 and 14 days after you bred her, you can *palpate* your doe to check for developing embryos in her abdomen. To palpate your doe, hold the rabbit's shoulders with one hand and use your free hand to push up into her abdomen, right in front of the pelvis. While doing so, feel around for the embryos, which at this stage should be like firm (but not hard) little grapes. If you encounter something small and very hard during this process, it is probably just a fecal pellet and not an embryo, but it can be tricky at first to tell them apart. Palpating within the 10-to-14-day window is important; if you do so earlier, you won't be able to feel anything because the embryos will be too small, and later on they become so big and soft that they get harder to distinguish from all the other important stuff inside a rabbit's abdomen.

I'll be honest with you—I'm terrible at palpating, and so are most of the other rabbit farmers in my network. However, mastering this skill is the only way to know for certain whether your doe is pregnant, so I suggest you try. But if for some reason you just can't get a knack for it, *test breeding* is your second best course of action for detecting pregnancy.

Test breeding is when you take the doe and put her back with a buck two weeks after the initial breeding. If she readily accepts a buck's advance, it is a good indication that the first breeding was unsuccessful, and you should adjust your calendar to reflect this new pairing. If she refuses the buck, it could mean she's pregnant *or* it could just mean she's not in heat. This is why palpating is more effective than test breeding. However, if you know your does' personalities well, test breeding can actually be pretty accurate. For us, the ways in which our does reject a buck when they are pregnant is very different from the ways in which they do when they are in diestrus. Our pregnant does tend to growl and become aggressive toward our poor boys, while our does in diestrus generally just ignore or run from them.

It has been said in some publications that test breeding should be avoided because female rabbits have what are known as *duplex uteri*. Females from species with duplex uteri have two uterine horns and two cervixes. This fact leads some to reason that it is possible for a doe to become pregnant with two litters, in different stages, at the same time. However, it should be noted that this phenomenon, called *superfetation*, is actually characteristic of female hares, and not of domestic rabbits. While I have read about a documented case of double pregnancy occurring in a domestic rabbit, it is exceedingly rare and should not be considered a serious risk for your herd. Rather, the real danger in test breeding is that your now angry doe will tear up your unsuspecting buck. So supervise your test breeds closely.

Gestation and Birth

Rabbits typically gestate for 31 days, although they have been documented to give birth anywhere between 28 and 35 days after being bred. The act of giving birth is called *kindling*. We have never had a doe kindle early at Letterbox, but we do experience the occasional overdue rabbit.

Preparing your doe for her babies' birthday is simple enough. At 28 days after breeding, separate your pregnant doe from other company and give her a *nesting box*. It can be stuffed with any combination of hay, straw, or pine shavings. When we used to make hay, we had a lot of it around so we used that. Now that we have to buy in material, we usually use straw. Straw has better insulating capabilities because it's hollow inside the stem. It's also drier and less tasty than hay, so the does don't accidentally eat it all before realizing it's for their nests—a practice we saw happen with the hay. While you don't have to wait until exactly day 28, do not put your boxes in too early, or the does might use them as bathrooms and soil the bedding before the kits are born.

The Nesting Box

A nesting box is exactly what it sounds like: a small box for a doe to build her nest in. Domesticated rabbits need us to replicate the shallow underground burrows their wild ancestors used for nurseries. These dens are just a tiny bit bigger than the rabbits themselves and have only enough room for turning around. Wild rabbits insulate them with soft grasses, leaves, and their own fur in order to keep them cozy. Rabbit farmers replicate these conditions using boxes lined with straw.

At our farm, we make our nesting boxes out of wood in order to save money, but you can purchase ready-to-use metal boxes from farm suppliers. The metal ones last longer and are easier to disinfect, but the wooden ones are a little warmer in the wintertime. Nesting boxes should be big enough for a doe to get in and out easily, but small enough to discourage her living in it.

We use untreated ⅔-inch (1.6 cm) plywood for our nesting boxes and secure the pieces together using 1¼-inch (3.2 cm) staples. (While

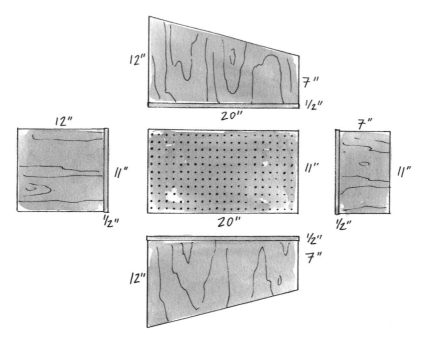

We build our nesting boxes out of plywood, using the specs shown in the illustration. You can also purchase them ready-made.

untreated wood does not last as long as the pressure-treated stuff, the latter is full of chemicals, and rabbits love to chew on the boxes.) For hardware, you can also use nails or screws. If you use screws, pre-drill your holes to avoid splitting the wood. Some nesting box plans call for hardware cloth on the bottom, rather than wood. In our experience, hardware cloth is easier to clean but can lead to frozen kits on cold nights due the lack of insulation. Unless your rabbitry is temperature-controlled or you live somewhere that is warm year-round, I would opt for solid-bottomed boxes. To clean your nesting boxes after each use, scrape out any material and submerge the box in (or spray generously with) a diluted bleach-water solution. Let them air-dry in the sun before using them again.

Kindling in the Nesting Box

Sometime between day 28 and kindling, your doe will shift around the contents of her box to make her ideal nest. This may involve taking out everything you have lovingly put into the box just to put it all back in the way she wants it. Don't be alarmed if this happens, because the old adage "mother knows best" is usually right when it comes to a doe and her nesting box. Either just before or shortly after kindling, the doe will pull fur from her dewlap and use it to line the nest in order to keep her kits warm. Experienced mothers usually prepare their nests well in advance of kindling, but first-time moms can be caught off guard by their first litter. It's common for does to line their nests after giving birth, so don't worry if any of yours seem behind schedule.

After we separate the doe and give her a nesting box, we let nature take its course. After all, there's really not a lot else to do. Rabbits tend to kindle in the evening, after the sun goes down, and the whole birthing process only takes about 15 minutes. It's pretty rare that we get to see it take place, so our protocol is to simply check on our expectant doe during morning chores on day 32. If everything has gone according to plan, we encounter a regular doe, looking like nothing's changed, hanging out near a box full of fur.

Aside from building a nest and lining it, a doe does not do much else to care for her young. She nurses twice a day and occasionally defends access to her space, but beyond those three things, she does not do a whole lot of active parenting. Unlike other newborn live-stock, such as lambs, calves, or chicks, newborn rabbits are basically

THE INSULATING POWER OF RABBIT FUR

The insulating properties of rabbit fur and hay never cease to amaze me. Our stockhouse is unheated, and while during the daytime temperatures are mild inside, the nights can get very, very cold. This year we had six of our does kindle during a night when it was 7°F (−14°C) out, and every kit survived.

A nesting box with newborn kits.

still in the fetal stage for the first week of their lives. They are naked and blind, which combined with a laissez-faire parenting style leaves the onus of keeping kits alive on us. Thankfully, the key to their vitality is only twofold: ensure proper nesting box design and utilize good hygiene.

Once you see that your doe has kindled, you can take a look inside the box and see how she did. Ignore the urban legend that a mother rabbit will abandon her babies if they have been touched by a human—it's just not true. Hopefully you will find eight or more warm, squirming, naked little babies. On the first day, I usually just fish my hand around the box to count the litter and remove any soiled bedding

THE IMPORTANCE OF A GOOD NESTING BOX

If the lip height for the opening of your nesting box is too low, baby rabbits will be able to flop out of their warm abode before they are ready. If this happens, even the best doe won't pick them up and tuck them back inside, and the kits can quickly die of exposure. Additionally, be sure to remember to fill the box up with clean substrate before you give it to the doe.

If a nesting box or larger environment is moldy, damp, or dirty, kits can develop inflammation in their nasal passages, dulling their sense of smell. Young kits rely on their sense of smell in order to find their mother's milk before their eyes open, so an impaired sense of smell likewise means impaired nutrition. In short, keep your nesting boxes clean and dry.

Kits are born blind and without fur.

These kits are nearly ready to lose the box.

or dead kits. Then I record the litter and leave everyone alone until the next day.

If you find a dead kit or two in your litters, don't freak out—it's normal. The United Nations conducted a study using exemplary French rabbitries in which they found on average that 5 to 7 percent of kits are stillborn, and 16 to 20 percent perish before weaning.[2] So long as your average kit mortality stays below 20 percent, there should be no cause for alarm.

At our farm we anticipate a 15 percent loss of kits within the first two weeks and do not worry unless we exceed this threshold or

Table 6.1. Kindling Quick Tips

What We Look For	Potential Problem	Solution
A nice layer of fur inside the box.	A poorly lined nest resulting in exposed kits.	Save a bag of clean fur from nesting boxes that are no longer in use to add to a poorly lined nest when necessary.
Movement from kits under the fur and bedding.	No noticeable movement could indicate stillborn, frozen, or very cold kits.	Remove the fur and bedding to see if the kits are alive. If alive but looking sluggish and feeling cool to the touch, warm them up immediately with a gentle source, like a hair dryer set to low or your own body heat. If the kits froze or were stillborn, remove them from the box.
Kits outside the nesting box.	Kits outside the nesting box can freeze, get trampled, or starve to death.	Make sure they are warm enough and tuck them back into the box. If the mom appears to have rejected them, use another doe with new kits as a foster.

something else seems unusual. If kits continue to die after two weeks of age, however, it could be an indicator of illness or poor hygiene, and action should be taken to correct the problem.

After your doe kindles, free-feed her and her kits until they are weaned. When you choose to wean your rabbits is up to you, but it's usually safe to do so after four weeks. We typically wean after six weeks, since we want our rabbits a little bigger before they go out into their pasture pens. You can remove the nesting box from the cage when the kits are big enough to jump in and out on their own accord.

Nursing

Rabbit milk has a whopping 12 percent fat content (that's more than three times the amount in cow or goat milk), and for this reason does need to nurse their kits only once or twice a day to keep them healthy and growing.[3] So if you notice that your doe never seems to be in her nesting box, do not panic. It's actually a rare treat to see a new mom nursing her kits.

Cannibalism

Once in a while, you might arrive at your rabbitry to find that a doe has seemingly eaten her own offspring. While cannibalism is rare, a doe will eat a kit that has died on its own but is still warm. Since rabbits are prey animals, instinctively concerned with staying on the down low, does may be eating their dead kits simply to get rid of any predator-attracting evidence. Hungry or malnourished rabbits may be more inclined to eat dead kits. Occasionally in our rabbitry we find a kit that appears to have been chewed at the belly or is missing extremities like ears, tail, or feet. This is likely the unintentional result of an inexperienced mom trying to clean off the afterbirth on her babies and taking it a little too far. While we have successfully raised earless bunnies, the kits that are missing limbs or other important body parts should be culled immediately to prevent suffering. Does that repeatedly chew their offspring should also be culled.

False Pregnancies

If your doe goes through the act of making a nest but never kindles, she likely experienced a false pregnancy. Ovulation and hormonal triggers can take place in a doe even if she was not successfully bred, causing her to go through the motions of preparing for a litter that will never arrive. For whatever reason, false pregnancies are a somewhat common occurrence in our herd. If you find your doe has had one, just remove the nesting box and breed her again.

Fostering

While does do not typically "abandon" their litter, sometimes the conditions just aren't right for a new mom to care for her kits. She may not be letting down enough milk to adequately feed them all, for example. Or if she's a new mom, she may just not have the right natural instincts. In any case it's always a good idea to plan for multiple does to kindle at the same time so there are foster mothers available for babies who are not receiving adequate care. Abandoned kits can be safely tucked into other litters, so long as the age gap between them is less than two days.

Breeding Schedule

Technically, a doe can be bred again almost immediately after giving birth. This means rabbits have the potential to rear almost 100 babies in a single calendar year.[4] However, just because it's possible does not mean it's a good idea. Overaggressive breeding will most certainly wear out your rabbits and compromise their health. However, breeding too conservatively can have a serious impact on economic viability. Table 6.2 assumes each doe produces six viable kits per litter, with each being sold for an average price of $25.

On our farm we breed our does 28 days after kindling. This schedule gives our animals adequate rest between litters and gives the kits

Table 6.2. Theoretical Rates of Reproduction and Correlating Returns

Days After Kindling	Litters per Year	Fryers per Year	Gross Income per Year
42	5	30	$750
35	5½	33	$825
28	6	36	$900
21	7	42	$1,050
14	8	48	$1,200

plenty of time to wean naturally. If a doe has an especially small litter—four kits or fewer—we might breed her a few days to a week earlier than usual next time in an effort to make up some lost time. Since she is rearing fewer kits, breeding a little earlier should not be a problem.

While they have the potential to be productive for a number of years, reproductive activity begins to drop around the age of three. We replace our bucks and does when they reach this age or earlier, if they cease to be productive.

CHAPTER 7

Record Keeping

I am not, by nature, a particularly organized person. I can *seem* organized to other people because I'm good at getting to appointments on time, always respond to my emails, and generally keep my spaces tidy, but boy oh boy do I lose records and receipts. For the first few years (yes, years—don't judge me) of running a business, I had virtually no production records and only the most rudimentary of income and expense reports. We were selling everything we produced, but we weren't making any money.

Because I had not collected any data, I couldn't figure out where the problems were. Was production too low? Too high? Was our pricing strategy wrong? Were we losing too many kits? Were our litters too small? How many animals did we have to raise and sell to meet our goal? There were so many questions, and literally no way to answer them. Laszlo, despite his ability to fix any piece of broken equipment, build something spectacular with $20 and a stick of gum, or get virtually any vehicle unstuck from the mud at any time, wasn't any better at keeping records than I was. Thankfully, it wouldn't be long before we joined forces with Faith, a vegetable farmer, my childhood best friend, and, conveniently for us, a real spreadsheet nerd.

Through our shared Google drive, I began to take note of all the things Faith kept track of on her side of the business and then all of the exciting information you could glean from it. I learned how to do things like categorize expenses and determine profit margins. I finally knew exactly how much time and money I had put into each enterprise. I knew what our feed conversion ratios were for our individual types of livestock. I knew exactly how much we were spending on ice. All of these things allowed me to make better decisions, and now each year not only do our margins improve, but so does our quality of life. Good record keeping has helped us work less and accomplish more.

Keeping Production Records

Depending on the scale of your rabbitry, production records can range from very simple to pretty complex. At a minimum, you will need to know:

1. Who your individual breeding rabbits are.
2. The *service date* and *service buck* (when they are bred and with whom).
3. The kindling date.
4. The litter size, noting live kits and number of stillborn and deceased.
5. When to breed again.

Noting just these five things will keep a small rabbitry running fairly efficiently. With this information, you'll be able to keep your herd productive by adhering to a regular breeding schedule. You'll also always know when to give a doe a nesting box and will be able to track the quality of your breeding stock by knowing how many viable offspring each individual produces annually.

Keeping this basic information in the wire-pasture hybrid system is simple. Each breeding rabbit has a name, and each cage has a tag with that name. When a doe is bred, the date and the sire (the buck used to service the doe) are recorded. Then, 31 days later, the litter is recorded. Any losses are noted, and the records are updated to reflect them.

We currently keep analog records in the stockhouse, so that any member of our team can take care of the rabbits on any given day, regardless of cell phone or computer access. Our analog records are super simple and look a lot like table 7.1.

Whoever is doing chores in the morning takes a look at the record book to see who needs to be bred or who needs a nesting box. If they gave a doe a box, then they check off the "Box In?" column so the next person knows it's taken care of. Then they check all the existing litters and update any changes in the notes section. (A quick tip for keeping perfect records: Tie a pen to your notebook, so it's always there when you or anyone else needs it. Farms are complex, busy places, and it's really easy to forget to return to small tasks after your brain has been baking in the sun all day.)

This analog data can then be added to a digital spreadsheet or a breeding software program. We currently use one called Hutch, made by BarnTrax. It's great because it's mobile-ready, so you can update your records directly from the field. Hutch automatically updates an

Table 7.1. Sample Breeding Records

Doe	Service Date	Service Buck	Nest Box Date	Box In?	Litter Size and Notes
Jezabel	Dec. 1	Mel	Dec. 29	x	Jan. 1: 9 live, 2 stillborn Jan. 3: 7 live, 2 dead
Donna	Dec. 3	Bruns	Dec. 31	x	Jan. 3: Doe made a nest, but did not kindle Jan. 4: No kits, false pregnancy, add to breed schedule
Carla	Dec. 9	Bruns	Jan. 5		

online calendar that tells the user when to put in a nesting box, when litters are due, and when to breed again. It generates cage cards with QR scan codes for easy date input and gives you reminders to track litter weights and suggestions on when your rabbits are ready to butcher. Most important, Hutch organizes all your data, so you can easily see how each doe or buck is performing. Hutch isn't free, but at the time of writing a deluxe membership is only $40 a year. If that's too much, I recommend using KinTraks, which is a free program similar to Hutch but with fewer bells and whistles.

At our farm we don't currently tag our growing rabbits in order to keep track of them once they are weaned. However, tagging can be useful for tracking performance. If you want to monitor your growing rabbits, you can tag each one with a permanent marker (you'll need to reapply regularly) or, even better, with a small tattoo inside their ear. Giving a rabbit a permanent ear tag is easy so long as you have a pair of tattoo pliers. They cost between $50 and $75 and the Ohio State University Extension Office has a great resource for how to use them, which you can access online for free.[1]

Slaughter Records

Because of the nature of our marketing plan, we process our meat birds and rabbits every Wednesday during the farm season. To keep track of our orders and have a record of every animal we bring to the slaughterhouse, I maintain a notebook in our stockhouse, which has been replicated in table 7.2. On Tuesday I write in the orders, so that the crew members who load them early Wednesday morning know what to do, but these notes also serve as an integral piece of our production records. Since we typically shoot for 3- to 4-pound (1.4–1.8 kg) rabbits, you'll see there is a permanent note reminding us of the correct live weight to ensure we meet our targets.

While it would behoove us to record the exact weights of each rabbit on the slaughter date, I would be lying if I said we did that on our farm. We just don't have the time in our mornings as it stands. However, as we gain efficiencies elsewhere, we will certainly be allocating the newfound time to keeping better records.

Table 7.2. Sample Slaughter Records

Date	Chix Retail Live Weight > 6#	Chix Wholesale Live Weight 4.5–5#	Rabbits Live Weight > 5.5#	Other	Order Filled (Y/N)	Notes
5/11	45	25	12		N	Only 10 rabbits were big enough
5/18	60	15	10	5 chix > 8# Label GFFN*	Y	

* This text is shorthand for "load 5 chickens who weigh 8 pounds or more, and label the crate with the account it's for (in this example, GFFN [Good Food Farmers Network])."

Sales Records

Our production and slaughter records let us know how many rabbits we have raised for market, so the next part is keeping track of what gets *sold*. We use a combination of Google spreadsheets and QuickBooks in order to do this, but there are other programs to choose from if you don't like these.

Since we sell all our rabbits by pound and not by unit, our sales records are where we glean information about our average finishing weights. Our market, wholesale, and CSA records all note the weight of each rabbit order, so we simply add all the weights together to get our total for the season. Once we have the final weight, we can use our slaughter records or the receipts from our processor to identify how many rabbits we processed in total. Divide the weight by the number processed, and voilà—we now know the average weight of our rabbits.

Expense Records

Once you know what you have produced and what you have sold, the final piece to the puzzle is knowing how much you have spent. Hutch has a nifty expense tracking feature, but here again we use QuickBooks. We synthesize the *operating expenses*, or the costs incurred during normal day-to-day operation, for our rabbitry into five categories: livestock, feed, processing, labor, and health/incidentals.

Table 7.3. Sample Expense Records

Item	Cost
Livestock	$0.00
Feed	$1,825.00
Processing	$918.00
Labor	$1,725.25
Health/incidentals	$250.00
Total costs	**$4,718.25**

Our *capital expenses*, or long-term investments, are categorized differently, but you can add a line for depreciation into your operating expenses if you would like to organize your information that way.

Putting the Pieces Together

Now that you have production, slaughter, sales, and expense records, you can figure out a lot of cool things about your rabbitry. For example, you now have all the information you need to determine your cost of production per rabbit, start-to-finish. To determine this, simply divide your total expenses by the number of rabbits produced. On our farm, our cost per rabbit is currently $14.20. Knowing how much something costs you to produce allows you to set your pricing appropriately. We will look at some other calculations later on in chapter 12 when we discuss profitability in more detail.

CHAPTER 8

Feeding Your Rabbits

The nice thing about our rabbits versus our poultry, for example, is that they need to be fed only once per day. We feed them in the morning, but our herd tends to snack a bit only while the sun is up and then finish off the rest in the evening. If we have full tractors or cages with large litters who are finishing up their food early, we may occasionally need to fill up their feeders again in the afternoon, but usually only on extra-chilly days when the rabbits need more energy to keep warm. As with all livestock, there are several different ways you can feed your rabbits. In this chapter I will provide information on our practices as well as an overview of rabbit feed basics.

Feed is the biggest expense in our rabbitry. This year rabbit pellets (pelletized feed made from a mix of grasses, grains, legumes, and minerals) made up a little over a third of our total expenses for this enterprise. Our cumulative annual feed bill across all enterprises accounted for 40 percent of our total livestock expenses, so it's important that we keep a close watch on what we are spending on it. Remember, this is a low-margin enterprise.

We use a 16 to 18 percent protein pelleted feed on our farm. No one on our team is a trained animal nutritionist, so we are careful to partner with mills we trust to take good care of us and our livestock. There are a lot of different ways to feed your rabbits, and different

CONSIDERING YOUR FEED

At Letterbox Farm, we use organic and non-GMO feed whenever possible. We are very fortunate to have Stone House Grain, a regenerative-focused, carbon-positive, affordable feed mill just 5 miles (8 km) down the road. This is where we purchase Hudson Valley–grown, Non-GMO Verified feed for our meat birds, laying hens, and pigs. However, there are simply not enough rabbitries in our area for this mill to produce organic rabbit pellets at a price that fits within our margins, and our current accounts are unwilling to absorb a higher price per pound to make up the difference. It is our hope that as the market for raising rabbits increases, so will the availability of organic and non-GMO feeds that are as affordable as that for other livestock.

If using organic feed is a top priority, be sure to find your mill so you can work out a recipe and negotiate a price before you begin your operation. In our area organic rabbit pellets are nearly $800 per ton, which is double what we pay at our conventional mill. Since the projections in this guide are based on feed that is $400 per ton, you will also need to adjust your numbers accordingly. From there, you can make a decision as to whether or not you can (a) increase the price of your product, (b) operate with lower margins, or (c) cut costs elsewhere.

mills will have different recipes. Don't be afraid to shop around, talk to other producers, or ask questions.

Cecotrophy in Rabbits

Because rabbits are herbivores, their plant-based diets contain a lot of cellulose. This insoluble substance is difficult to digest for all animals, but herbivores have specific ways to cope with the challenge. Ruminants, like cattle, sheep, and goats, have special four-chambered stomachs to get the job done. The first chamber is the rumen—hence the category *ruminants*—which contains a salty solution full of specialized bacteria. These bacteria work over time to break down plant material via fermentation, but the animals still have to regurgitate the partially broken-down plant material, called cud, and chew on it some more to further break down the cellulose. This cycle repeats until the cellulose is finally broken down enough to pass through the rest of the digestive tract. Ruminants are called *foregut fermenters* because this whole process takes place before any food enters the small intestine.

Rabbits, on the other hand, are *monogastric, hindgut fermenters*. They have only one stomach and no rumen, so for them the microbial fermentation necessary to digest cellulose takes place after the small intestine, in the colon and *cecum*. After rabbits ingest and chew their food, it moves to their stomach, where it sits for a few hours to be partially broken down by acids. Next it moves into the small intestine, where it's broken down even more. From there most of the digestible particles are absorbed, while the indigestible fiber moves along into the colon. This is when and where those hard and round little poops are formed. The remaining digestible material, however, moves on into the *cecum*, which is the largest organ in a rabbit's digestive system. Like a rumen, the cecum is full of symbiotic bacteria that ferment whatever is left, after which some of the contents are ready to be absorbed. Whatever is not yet broken down enough to be absorbed, however, is expelled into the colon and formed into soft pellets. The soft pellets, called *cecotrophs*, are expelled through a rabbit's anus just like excrement. However, unlike the hard feces, rabbits actually ingest the mushy cecotrophs so they can run them through the whole digestive system again and absorb even more nutrients. Some fractions of a rabbit's meal will go through this process as many as four times.[1]

Cecotrophs are consumed directly from the anus, usually at night in domestic rabbits. According to Amy E. Halls, a monogastric nutritionist for a large feed supply company, healthy rabbits will consume most or all of their cecotrophs, while sick rabbits may leave large amounts behind with the rest of the poop. You can differentiate cecotrophs from regular excrement because the pellets are much smaller and softer. They are also very shiny and form clusters, unlike regular pellets, which are dull and singular.[2]

The magical process of cecotrophy is what allows rabbits to consume poor-quality, high-fiber diets and still obtain all the necessary nutrients. It's a nifty little feature that's found only in rabbits and their cute little cousins, pikas.

What Does It Mean to Be "Grass-Fed"?

Since rabbits are herbivores, they can be 100 percent grass-fed just like cattle, sheep, and goats. Despite the fact that grass-fed rabbit was once ubiquitous throughout Europe, today this type of production is rare, though there are still certainly some compelling reasons for raising rabbits in this tradition. For starters, it's natural. Rabbits have evolved on forage for more than 40 million years. This means they know exactly what to do with all that vegetation out there. When given the opportunity, even domestic rabbits will instinctively create a perfectly balanced diet for themselves—picking and choosing exactly what they need from the environment around them.

Second, depending on the size and nature of the farmscape, it's free. Studies have shown that, with a decently healthy pasture, it's possible to grass-feed and finish up to 60 rabbits per acre (150 per ha), annually.[3] This means that if you've got 20 acres (8 ha) of arable land at your disposal, you could, in theory, raise 1,200 rabbits per year without spending a dime on feed. Grass-feeding your rabbits could also give you a leg up over other rabbit farmers when it comes to marketing your product. When given the choice, Americans are increasingly opting for meats with that grass-fed label,[4] and while pasture-raised rabbit doesn't currently fetch much higher a price than conventionally raised, it certainly could in the future.

And yet, despite all this, on our farm we opt for a pelleted feed. If you're wondering why, remember this: Just because something is *possible* doesn't mean it *works*—well, not reliably enough for a commercial farm, at least. A common factor present in all the reports on grass-fed and finished rabbit that we've come across is slow growth—really slow. The best-performing cage-raised, conventionally fed rabbits can reach a dress weight of 3 pounds (1.4 kg) in as little eight weeks. Rabbits raised in our pasture-wire hybrid system at Letterbox reach that same weight between 12 and 16 weeks, quite a bit longer. Now compare this with rabbits that are 100 percent grass-fed. It takes them around six months to reach that very same 3-pound dress weight. That's twice as long as a well-performing Letterbox Farm rabbit and more than three times as long as their conventionally raised counterparts.

Right now, your rational brain might be wondering: *How does any of that matter if the feed is free?* It's true that a cost of $0 per week is still $0 whether you multiply it by 8 weeks, 12 weeks, or even 25. But labor isn't free, and the longer it takes to get a rabbit to market, the more expensive it gets. More important, though, the longer an animal takes to reach market weight, the more opportunity there is for something to go wrong.

Grass-fed rabbits raised on pasture require large, open pens, like the ones used in the Coney Garth method discussed earlier in this book, in order to obtain the proper nutrition. This increases opportunities for predation, escape, injury, and contact with wild organisms, which can transmit disease and cause illness. The ratcheted-up risk combined with the vastly extended time line has proven to be a difficult combination for commercial producers. Remember, the farmer who developed the modern-day Coney Garth system of management reported a 70 percent loss rate. Losses that high just aren't sustainable in commercial production.

You could, however, grass-feed rabbits in a less risky setting by keeping your breeding rabbits, or even your whole herd, indoors and feeding them hay. This amendment could very well drastically reduce the rate of loss, but unfortunately it doesn't much speed up the production time line. This is problematic even beyond the issue of increased labor, particularly in the United States, where the vast majority of the rabbit market is for young rabbits, known in the industry as fryers. In order to be considered a fryer, a rabbit has to be small (under 3½ pounds / 1.6 kg, dressed) and it has to be young. By the time a rabbit reaches six months old, it's developed a different flavor and texture, which moves it from the fryer range and into the *roaster* category. Roasters, while popular and even preferable in many places abroad, Europe especially, just aren't very marketable here in America at the moment. While culinary trends can always change, right now the vast majority of us like rabbits young, tender, and mild.

Raising our grow-outs on pasture is mostly a choice we made to increase animal welfare in our rabbitry. We do it because it keeps our rabbits active, engaged, and healthy, without putting them at too much risk. We also do it because it builds soil, increases fertility in our fields, reduces pressure on our limited barn space, and makes our rabbits taste better.

One day roasters may finally have their breakout moment in the American culinary scene (someone was just telling me about the growing popularity of *vaca vieja* in haute cuisine last week, so you never know). Maybe then we'll dive back into researching and implementing a 100 percent grass-fed rabbit production. In the meantime we're more than happy to support our local feed mill, which thankfully makes a great well-balanced and affordable feed that our rabbits think pairs quite nicely with the forage they find inside their tractors.

How Much to Feed

How you should feed your rabbits depends on how big they are and what stage of their life they are in. In her book *The Rabbit Raising Problem Solver*, rabbit expert Karen Patry breaks down ration guidelines into four phases of a rabbit's life: adult bucks and adult dry does, juniors (from weaning to adult), pregnant does, and lactating does.[5]

At our farm, for *adult bucks and adult, non-lactating does*, we use a baseline of ¾ ounce (21.3 g) of pellets for every pound (0.5 kg) of a full-grown rabbit's weight. So under normal circumstances, we feed a 10-pound (4.5 kg) rabbit 7 ounces (200 g) per day. *Juniors* or grow-outs, which are weaned and fattening rabbits, are free-fed as much as they want until slaughter or, if we are saving them for breeding stock, until they reach their adult size. For the latter group, we gradually reduce the feed for a couple of weeks while they adjust to their new rations.

Pregnant does are finicky eaters. About midway through their gestation, ours tend to get a bit hungrier than usual. During this time, we give them an additional 1 or 2 ounces (28.4–56.7 g) of pellets and

We fill up the feeders for our grow-outs.

maybe some extra hay or fresh greens to snack on. A few days before kindling, though, we often notice a sharp decrease in their appetites and adjust the rations accordingly to prevent pellets from building up in the feeders. *Lactating does* and their litters also get free feed. Patry recommends adding a couple of tablespoons of black oil sunflower seeds to aid in milk production, and we do the same.

None of these guidelines are hard-and-fast rules in our rabbitry. Rabbits are pretty good self-regulators of their own nutrition, so we make sure we listen to them when they send us signals. For example, if our animals aren't finishing the food we give them throughout the course of a day, we know we are giving them too much (or they are sick, but that's another chapter). If the feeders are empty and they seem really excited to see us, they probably need a little more feed. Unless your rabbitry is climate-controlled, you can expect dietary needs to ebb and flow as the weather changes.

Supplements

We use a nutritionally complete commercial feed for our rabbits, so we don't really need to amend their diet with much else in order for them to thrive. We supplement them anyway, though—mostly for their own entertainment, but also because a diversified diet is always good for health and, as an added bonus, it makes meat taste better.

When our grow-outs are out on pasture, we have pellets available and let them pick and choose their own treats from what's around them. I will sometimes toss some twigs from our fruit trees into the rabbit tractors because they are high in tannins, which aid in parasite control, and because their presence distracts the rabbits from chewing on the wood from which the tractor is constructed. It's kind of amazing how much damage they can do with those little teeth of theirs. Since our breeding stock is up on wire, we bring the treats to them. Some of our favorites include hay, comfrey, sunflower seeds, and twigs.

Hay

Low-grade timothy or meadow hay are both great for adding fiber to your rabbits' diet. You can also give your rabbits small amounts of alfalfa hay, but I don't recommend it for a few reasons. For one, at least where we are in the Hudson Valley, it is much more expensive and more difficult to source than regular old meadow hay. Alfalfa is also a main ingredient in rabbit pellets, so it doesn't add much diversity to their diet as a supplement. Last, when dried into hay, alfalfa becomes very rich and calorie-dense. This is fine for young and growing rabbits, but it is not good except in small quantities for mature ones who need to maintain a target weight to stay healthy.

We pretty much use only meadow hay on our farm, which consists of a variety of grasses as well as pieces of other plants and a few sticks here and there. It's inexpensive at $3.50 to $5.00 per bale, depending

Comfrey is easy to grow and is a great supplement for your rabbits. *Photo by Grahamphoto23.*

on the time of year, and is easy to find. We give it free-choice to our breeding stock, but make sure you don't put more in their cage than they can finish up in a day. Otherwise, they may poop and pee on it, spoiling the hay and dirtying the cage.

Comfrey

Comfrey is a perennial leafy green in the borage family. It's super easy to grow and is known for its usefulness as a medicinal herb for humans, pets, and livestock. We like it for rabbits because it's a good source of vitamin A, a nice digestion aid, and an overall health booster. Comfrey leaves can grow quite large, so one medium-sized leaf per bunny is plenty—too much may cause diarrhea. Comfrey can also cause liver damage if consumed in too large quantities.

Black Oil Sunflower Seeds

I've already mentioned these a few times, but to reiterate, sunflower seeds are great for body conditioning, especially in the winter. They also help lactating does with their milk production. To feed, put one small handful per rabbit into the feeder.

Twigs

Chewing on twigs and branches is good for your rabbits' teeth. While their regular diet of pellets is usually enough to keep their constantly growing chompers in good condition, giving your rabbits some fresh wood once in a while can keep them appropriately worn down. Choose varieties that are high in tannins, like willow, hazelnut, oak, ash, and pear, to help prevent coccidiosis.

Storing Feed

Rabbit pellets tend to spoil faster than other animal feeds, so be sure to store them somewhere dry and out of the sun. If you can, buy only one month's supply at a time, especially during the summer months. For us, this is conveniently 1 ton, or 40 bags (50 pounds/22.7 kg each) total. I find that stacking our feed on a pallet that has been elevated off the ground using cinder blocks, two high, helps significantly when it comes to keeping rodents away.

Water

As with all living things, the availability of fresh clean water is paramount. Ideally, your rabbits should have access to water 24 hours per day. This can get tricky during the colder months, as water lines, bottles, and bowls can freeze up in cooler climates. When this is the case, just be sure your rabbits have plentiful access to fresh water at least twice per day. Individual steel bowls or well-made plastic pen cups are a good option for this. Dishes that can hook onto the cage are better than freestanding containers, since even thirsty rabbits have an annoying habit of tipping over their dishes before they take a good drink, or, worse, pooping in it.

All of our rabbits live outside in an unheated greenhouse for the colder months. While it dips below freezing after the sun sets, the temperature heats back up during the day, allowing our plastic water lines to thaw out and function just fine by 10:00 AM on most days, even in midwinter. We have pen cups available for cloudy or extra-chilly days and often fill them up, especially during warm days in summer in an extra effort to ensure the rabbits drink enough.

How much your rabbits drink will vary wildly depending on the circumstances of the day. Rabbits out on pasture or who are fed a lot of fresh greens will drink less because they are getting metabolic water from the plants. They will drink less on a mild day and more on a hot day and, likewise, less when they are dry and more when they are lactating. To get a sense for an average, though, note that studies have shown full-grown rabbits to drink roughly 10 ounces (0.3 L) of water each per day.[6] Using this as our guideline, we can surmise that we will need approximately 1 gallon (3.8 L) of water for every 12 rabbits in our herd. Considering that smaller rabbits will drink proportionately less than their full-grown counterparts, this approximation is an overestimate, but when it comes to water, it is better to be safe than sorry.

While it is of utmost importance that your rabbits have regular access to water, don't panic if you forgot to check your water source one day and realize the next that it wasn't working. Rabbits can actually survive with no water at all for as long as eight days, if conditions aren't too hot or humid. So if for some reason your rabbits' water supply is restricted for a day or two, they won't suffer any long-term damage, if their access is promptly restored. Rabbits without water, however, won't eat and can experience quite a significant drop in weight in a short amount of time.

Rabbits are even more resistant to hunger than they are to thirst. They can survive for up to four weeks without feed (although probably not that comfortably), so long as they have lots of water instead. This, of course, is just a silly piece of trivia—you should definitely feed and give water to your bunnies every day of the year.

Signs of Dehydration

If it's winter and your water availability is sketchy, of if you just realized your water lines were clogged for an unknowable length of time, knowing the signs of dehydration is important for monitoring the health of your herd. Look for these three things:

1. **Good skin turgor.** To assess this, use two fingers to grasp the skin on the back of your rabbit's neck and hold it in a tented position for a few seconds. In well-hydrated animals, the skin will quickly snap back to its original position. If it takes a few seconds to settle back, your rabbit is dehydrated.
2. **Sunken appearance.** Your rabbits should look full and round. If parts of their body appear sunken or thin, your bunny may be dehydrated.
3. **Loss of appetite.** Rabbits will not eat without water. If they haven't eaten their rations, they may not be getting enough access to fluids.

If any of your rabbits seem dehydrated, make sure they have constant access to fresh water. To encourage drinking, you can add a splash of apple cider vinegar to their water supply since acidity stimulates salivary glands. This makes the act of drinking water feel more refreshing, in turn making rabbits want to do it even more.

The Cost of Feeding Your Rabbits

While feed conversion differs from farm to farm, you can use averages collected from multiple rabbitries by the Food and Agriculture Organization of the United Nations (FAO) in order to budget your annual feed costs before you have collected your own data. The data in table 8.1 were collected throughout Europe in rabbitries most commonly using Californian and New Zealand rabbits.[7]

When all these constantly shifting pieces are looked at together and averaged, this study concluded that the average rabbitry operating at a semi-intensive rate (35 kits per doe per year) uses 2¾ pounds (1.3 kg) of feed per breeding doe per day. I cross-checked this with our own production records, and this figure is actually pretty close. To estimate your annual feed cost, use the product of 2¾ (pounds per day [1.3 kg per day]) and 365 (days per year), or 1,003¾ (474.5) in the following formula:

Table 8.1. Feed Consumption by Life Stage

Stage	Notes	Feed Consumed per Day
Juveniles, grow-outs	4–11 weeks	3½–4 ounces/99.2–113.4 g
Lactating does with litters	Weaning at 4 weeks	12–13 ounces/340.2–368.5 g
Adult rabbits	Maintaining weight	4–4.5 ounces/113.4–127.6 g

(1,003.75 × feed price per pound) × number of does
(474.5 × feed price per kg) × number of does

So, using our feed price, which is $0.19 per lb., the annual feed bill for a 10-doe rabbitry producing 350 salable offspring would look something like:

(1,003.75 × 0.19) × 10 = **$1,907.12**

Contrary to popular belief, raising our rabbits partially on pasture does not reduce our feed costs as far as we can tell. The fact that our averages mirror those collected from more conventional rabbitries, despite our usage of free supplemental feed, leads me to believe that whatever the savings we gain from the available pasture, we lose due to the increased physical activity our rabbits exhibit in the tractors. Of course, the gains we make in increased animal welfare are well worth this dip in productivity in our eyes.

Feeding and Watering Equipment

We use 9½-inch-wide (24.1 cm) Siftomatic metal feeders from KW Cages in all of our rabbit housing. They are advertised as able to hold up to 68 ounces (1.9 kg) of feed, but I have never measured it to confirm. Whatever they hold, it's plenty, and at nearly 10 inches across, one is wide enough to serve up to 12 rabbits. Ours have the standard

Our watering system. *Photo by Nichki Carangelo.*

mouth opening of 2½ inches (6.4 cm), but feeders with extra-large openings of 3¼ inches (8.3 cm) are also available for any of your big-headed bunnies.

Installing this type of feeder in a standard wire cage requires cutting out a rectangle that is roughly ½ inch (1.3 cm) larger than the feeder all around. We install two of these feeders into each rabbit tractor by screwing them from the base into the center beam that runs down the middle. We put the feeders on the inside rather than hanging them onto the wire from the outside like we used to. This is for a couple of reasons. For starters, having them inside keeps the feed dry in inclement weather. It also keeps them safe from other hungry animals, like raccoons, birds, and, most important, wild rabbits, whose contact with our domestic herd could cause illness and disease. If you don't buy a Siftomatic feeder, I highly recommend that whatever you do purchase has the same perforated bottoms as theirs. The holes in the trough allow the dust that can build up from pelleted feeds to fall right through, which is important as rabbits are very sensitive to dust.

For waterers, we use regular 5-gallon (20 L) buckets fitted with ⁵⁄₁₆-inch (7.9 mm) (inner diameter) black plastic tubing that leads to one small-animal-style nipple waterer per cage. To install the tubing, drill a hole in the wall of the bucket, close to the base, that is slightly smaller in diameter than the tubing. Next, use a lighter or a torch to soften 2 inches (5.1 cm) at the end of your tubing. Stuff the now highly pliable end of the tubing into the hole using a pair of pliers to pull it through until all of the tubing that was softened with the torch is inside the bucket, plus another inch (2.5 cm) that was not softened. Trim off the soft part of the tube, and your bucket is ready to use.

We use one 5-gallon (20 L) bucket for every 10 cages in our rabbitry, even though 5 gallons of water could easily supply more rabbits than that. I have found that the water pressure can get a bit low with cages beyond that, unless your equipment is sloped down-hill (ours is level). Also, the tubing can sometimes get gummed up, which blocks water from the nipples. When this happens, having shorter lines makes it easier to locate the problem area and clean it out.

Considering Alternative Feeds

As I mentioned, we use a commercial pelleted feed to make sure our rabbits' dietary needs are met, and while I recommend you do the same, you do have other options. In fact, there are entire books dedicated to alternative feeds for rabbits. Interest in the subject is currently growing, particularly in parts of the world where the climate is not suitable for growing the ingredients typically used in commercial rabbit feed. With climate trends shifting so rapidly thanks to global warming, we truly do not know for sure what can be grown where in the future, so learning about alternatives now could be a good investment.

German trials in direct grazing, like we saw used in the Coney Garth system earlier, have demonstrated that rabbits grown in natural meadows can feasibly produce 200 pounds of protein per acre (224 kg / ha) per year.[8] Given that an average rabbit in the US market dresses around 3 pounds (1.4 kg), we can conclude that it is possible run a small rabbitry aimed at producing 300 rabbits per year with just 5 acres (2 ha) of marginal pasture. The German trials in this system did, however, note a similar reduction in growth rates to those seen in the Coney Garth report. Rabbits raised exclusively on forage gain weight at roughly half the rate of their hutch- and cage-raised counterparts, even though they consume a lot more feed. This fact, coupled with the higher risk of predation, escape, and disease, makes this method of running a commercial rabbitry challenging to execute, despite being theoretically possible.

A lot of other great studies have been done over the years testing out alternative ingredients for feed in production rabbitries. Here, I will summarize some of my favorites from a list primarily curated by the FAO.[9]

Beets. Some old-school European rabbit farmers use fodder sugar beets to feed their animals, particularly in the winter months. Beets contain 17 to 18 percent protein, and both the leaves and the roots are digestible for rabbits. They are, however, very rich in minerals, which may cause digestive issues.

Wild carrots. Endemic to Europe and Southeast Asia, wild carrots are also naturalized in the United States. Here, we refer to them more commonly as Queen Anne's lace. Wild carrots are another traditional feed for Old World rabbits, but they are also being used as fodder in tropical countries and in Africa thanks to their high tolerance for variation in climate and soil.

Sweet potatoes. While these are predominantly grown for human consumption, surplus sweet potato crops have been used as feed for rabbits in Mauritius, Guadeloupe, and Martinique. Rabbits can eat the roots, which have lots of energy thanks to the 70 percent starch content, as well as the tops, which are considered highly digestible. Tubers can be chopped up, dried into chips, or ground into meal.[10]

Mulberry leaves. In places where mulberry leaves are not being used to feed silkworms, rabbit farmers may choose to use them as their main source of nutrition. Adult rabbits can actually subsist on mulberry leaves alone, which is pretty neat.

Poplar leaves. Trials in the United States have documented that green poplar leaves can replace dried alfalfa to effectively make up as much as 40 percent of a rabbit's diet. Leaves from new growth are preferred over those from mature branches, as they contain more protein.

Sugarcane. Farmers in parts of the world with wet, tropical climates have successfully used sugarcane to augment balanced feed rations. A trial in Mauritius proved that chopped-up sugarcane can replace up to 50 percent of a growing rabbit's diet, with zero drop in growth and performance. When given a choice, rabbits in another study replaced 40 percent of their balanced rations with sugarcane all on their own. Rabbits showed a preference for dry leaves, then green leaves, and lastly the coarsely chopped cane.

Cooked potatoes. While it is fairly well known that raw potatoes aren't usually good for livestock, rabbits can actually eat cooked potatoes without issue. It doesn't make much sense to put rabbits in direct competition for human food, but it's interesting to know that backyard producers can give their rabbits some common household waste products, like potato skins. Again, they need to be cooked, and it is important to avoid any and all green parts of the potato even so. These parts, which result from premature exposure to light, are toxic.

Coconuts. Some producers in Martinique and Sri Lanka have started feeding coconuts to their rabbits. One study showed that they could even form up to 30 percent of their total diet. Apparently, rabbits enjoy the green meat that's left after all the milk has been consumed.

Other by-products. Some researchers have looked into the possibilities for feeding by-products in commercial rabbitries. This is of particular interest for use in regions of the world where there is significant competition between livestock and humans for high-quality foods. In Côte d'Ivoire rabbits are sometimes fed waste from the pineapple canneries, although not in high quantities because they are very low in protein. In Burkina Faso spent grain collected from breweries has also been used to feed rabbits. What is notable here is that this by-product of beer can be used to make up as much as 80 percent of a rabbit's diet, so long as the additional 20 percent is properly composed. When mixed at this ratio, rabbits fed augmented spent grain even outperformed rabbits raised on a balanced commercial feed!

At our farm we have the luxury of simply calling up our local feed mill, which always has in stock whatever we need. Still, I am interested to learn about what's happening in other parts of the world for a couple of reasons. Some of the most important innovations in agriculture have emerged from conditions where access to resources is restricted. Also, farms in less industrialized nations often look a lot more like a small integrated farm such as ours than the average large-scale commercial farm in the United States. We might have a lot more in common with farmers in Burkina Faso than we think—after all, there are an awful lot of breweries in the Hudson Valley.

An Argument Against Medicated Feeds

In your search for the perfect rabbit pellet, you may stumble across medicated feeds. A medicated feed is a mixture of large quantities of feed and a veterinary premix of medicinal products. Most commonly, they contain coccidiostats intended to prevent parasites before they occur. While we use antibiotics and other medications to treat sick animals when absolutely necessary, we do not use any medicated feeds on our farm. This is a personal choice we made because as a business, we do not believe in the practice of using medications, especially antibiotics, preventively. The invention of these products is one of humankind's greatest accomplishments, but their frequent and inappropriate use can cause bacteria or other microbes to change, rendering the miracle drugs ineffective. We don't want to contribute to the problem of medicine resistance if we don't absolutely have to.

If you do choose to use a medicated feed, be sure to take note of the withdrawal time on the label, as it is illegal to sell (and icky to eat) meat from animals that have not met a specific withdrawal period before slaughter. While they do not have guidelines for rabbits, the Animal Welfare Approved regulations require that certified producers double any recommended withdrawal periods in order to meet their standards.

CHAPTER 9

Health and Disease

When I talk to non-farmers about my work, people often assume that bringing animals to slaughter is the most difficult part of my job. They're right in assuming that I definitely still struggle with the task of taking the life of another living thing, but it's really not the thing that keeps me up at night. For me, the hardest part of being a livestock farmer is being responsible for the health and well-being of every animal on our farm, every day. Dealing with unintended death and disease is heartbreaking, and staying focused and positive when things are going wrong is still the hardest part of my job. I remember reading a popular guidebook for raising meat birds back when I first started farming. Laszlo and I were pretty much following it like it was the Bible, so I read that book closely. In a chapter about sickness and disease, the author said something along the lines of, "If your animals get sick it's your fault—just do a good job, and they won't get sick." *That's easy enough*, I remember thinking. *I'll just do everything right and then I'll never have to deal with any sick or injured animals!*

Our first batch of chicks arrived, and I diligently followed instructions and tucked each tiny beak into water, carefully watching for a nice big gulp before gently placing the baby birds one at a time into a brooder built exactly to author's specifications.

The first few weeks went amazingly well, with the exception of the time a visiting friend trampled two chicks, having not watched their tread inside the brooder. So far, the author was right—do everything correctly, and all will be fine. Do something stupid, like let someone stomp around blindly in a small room full of tiny birds, and problems will arise.

Right on schedule, we moved our young birds from the brooder to the pasture early in the morning, just like the book suggested. For another week, we were in pastured poultry heaven. Perfect, happy birds growing fatter each day. Then, sometime in week six or seven, disaster struck. I arrived at the field feeling great, only to find a chicken lying there just splayed out on its back, dead. I freaked out. I cried. I scoured the internet for advice on what to do. I called the University of Connecticut's Cooperative Extension. I called Cornell University. I called random chicken farmers around the state desperately seeking to learn what egregious act of neglect caused this mysterious death. All I could think about was how, in ordering those chicks, I had made a tacit agreement with the universe that I would take the absolute best possible care of them until their one bad day arrived, and somehow, despite my best efforts, I had failed. I felt so guilty.

What I learned from that day from those strangers on the internet, the extension office, and, most helpfully, the more experienced farmers, can be summarized in three words: Sometimes animals die. They die or they get sick, even when you've done your absolute best to keep them alive and well. As bad as it may sound to the outside world, many animals have died under my care. Sometimes it's 100 percent been my fault, sometimes it's more like 50 or 25 percent. Plenty of other times, it's completely out of my control.

I am not sharing this fact because I want to encourage you to take a sloppy or laissez-faire approach to raising animals. Quite the contrary—I know that livestock require us farmers to be on our game all the time. But new farmers should know that sometimes, the reasons our animals fall ill or injured are complex. On top of it all, they can't tell you what's wrong—you just have to know or take your best guess. Sometimes identifying the problem is easy, but finding the solution is really, really hard. Other times the issue takes weeks or even months to diagnose but the solution is simple. To get good at raising animals, it takes time, experience, and the steady development of both knowledge and instinct—and even once you do all that, animals still die. Anyone who tells you differently is likely trying to make you feel better.

The key to dealing with this fact is learning to forgive yourself for your mistakes and then figuring out how to ensure you do not make them again. So long as you are really trying on that second part, there is no reason to be hard on yourself. All that does is slow you down. Somewhere in the top five best pieces of advice I have ever received is wisdom gleaned from my old mentor: "Things go wrong in farming every day. The key is to figure out how to sleep at night anyways." It sounds easy, but since sustainable agriculture appears to self-sort for people who are obsessed with having the best possible impact on the world around them, we farmers tend to dwell on our failures.

I've needed to lean into this advice pretty hard myself lately, as a matter of fact. This winter some of our juvenile rabbits began to die with almost no warning or obvious signs of illness. Jamie, our livestock crew member, and I performed necropsy after necropsy and while it was obvious our animals were dying due to intestinal blockages, we could not figure out what was causing the problem. We looked for potential environmental issues and couldn't spot a thing. We tried increasing fiber in the diet, we tried decreasing fiber in the diet. We used apple cider vinegar, probiotic powders, and coccidiostats—all to no avail.

It was clear that I needed a fresh set of eyes on scene—maybe another trained professional could spot something I was overlooking. So I called our extension agent, Ashley, and she came by right away. I was thrilled to hear her say how perfect she thought our rabbitry was—she even said it was the cleanest one she'd ever seen! It was a huge relief to learn I hadn't been overlooking something amateur, but at the same time it was disheartening to learn that even she couldn't identify the problem. Time was ticking away; rabbits were dying and nothing I could think of was stopping it. I needed to figure out how to sleep at night.

We sent some fecal samples to the lab to see what they could tell us about our situation, and with those results Jamie and I were able to piece the whole thing together. While, thankfully, most of the samples came back negative for everything, the ones from the sick and dying litters were positive for an obscure type of intestinal coccidiosis I had never seen before. It had different symptoms, which is why we couldn't really recognize it. Plus, this particular strain responds only to very early intervention with a coccidiostat, which is why our treatments weren't working.

I'm still not 100 percent certain how any of our rabbits caught exposure to this disease in the first place. If it had happened in the field, I'd simply blame the ground, but this nasty disease got into our wire cages, which we use for our breeding rabbits and their young kits in order to avoid *this very problem*. While I'll never know for sure, I do suspect it all traces back to one particular day when a hundred or so laying hens barged their way into the rabbitry (much to my dismay) and flapped around with their muddy feet right on top of our rabbit cages, compromising our very intentional biosecurity measures.

Ultimately, I know that at the end of the culpability train of events it's my fault the birds got into the rabbitry. But I also know that beating myself up for an honest mistake—that, if we're being real, would not have been an issue under 99 percent of circumstances, as many people successfully raise their chickens and rabbits together—is a pointless exercise. So I got my rest, stayed sharp, worked hard, and now I'm happy to say we are on our way to recovering from this challenging but also informative setback. I know it won't be long before we emerge from this experience as better, smarter farmers.

Now that you know that you are going to encounter some problem at some point inside your rabbitry, let's go ahead and start building that arsenal of knowledge you'll need to be an effective problem solver. Rule number one in animal health and well-being: *Prevention is the best medicine.* As the FAO's extensive guide to the rabbit notes, the strongest tool a producer has in cultivating healthy rabbits is the rabbit's own ability to keep disease at bay.[1] The guide goes on to explain that disease, pollution, and contamination are out there, lingering in the world around us (on a perfectly innocent chicken's foot, perhaps!). The key is to keep all that stuff away from your animals. The name of this game? Preventive hygiene.[2]

The Importance of Preventive Hygiene

Unclean waterers and feed troughs can foster harmful mold and bacteria. We scrub out our water buckets and flush out the lines every two to four weeks (more frequently, if needed) using a diluted bleach solution (approximately 1 part bleach to 10 parts water). This knocks back any bad stuff growing in there and gives us a chance to scrub or blow out the algae that love to bloom. Adding apple cider vinegar to your water source also helps keep your equipment clean in between. As I've mentioned, it's a good health supplement for your rabbits as well. Use 1 to 2 tablespoons per gallon (3.8 L) of water.

We use only sifter-type feeders in our rabbitry. They have small holes in the trough area, which allow dust from the pellets to fall through. Since rabbits don't eat the dust, making sure it has a place to go eliminates the buildup of old feed. This is important because old feed gets moldy, and mold causes illness. We clean our metal feeders by brushing them off, spraying them with a diluted bleach solution, rinsing them off with fresh water, and letting them dry before we put more feed in them.

Because we use metal, wire-bottomed cages for our breeding stock and open-bottomed tractors for our growers, there's never much debris or fecal matter in any of our housing (as there can be with wood and other solid-bottomed cages or indoor colony pens). We walk up and down with a wire brush to quickly brush out any pellets that have not fallen through the mesh on their own, and that's typically enough to keep our breeders' cages tidy. Our bucks' cages, for whatever reason, get really furry, so we hit them with a handheld blowtorch as needed to burn off whatever the rabbits have shed (obviously not when the rabbits are inside!). This keeps our cages looking nice—and because loose hair harbors germs, it's best to prevent it from building up. Flaming metal cages and equipment can also knock back viruses, parasites, and fleas, so we hit all of our cages once a quarter or anytime we have had any sign of potential outbreak. You can gently flame wooden hutches and obtain the same benefits—you just need to be more careful not to light them on fire.

We use the same trusty diluted bleach solution to clean our nesting boxes after each use. Our boxes are wooden, so it means they, like other wooden equipment, need to get good and saturated in order to be effectively disinfected. For this reason, I find it easiest to fill a tub with bleach water and submerge the boxes for a few seconds. After that we put them in a sunny spot for a couple of days to dry out before we use them again. Having some redundancy in your equipment is helpful here, because it allows you to rotate out, clean, and properly dry things without halting production. All of these steps become quick and routine once you are in the habit of performing them.

Keep in mind, you cannot disinfect dirty equipment. Always rinse or brush off any dirt beforehand and let everything thoroughly dry afterward.

Limit Outside Contact

Many of the more technical resources available for rabbit producers stress the importance of maintaining a *closed rabbitry*. This means restricting access to only the proprietors and trained staff. Technical advisers may also call for the regular use of hand sanitizers and the donning of sterile gear, like booties and smocks that shield your rabbitry from any potential disease vectors living on your street clothes. That United Nations guide that I love so much calls the risk of outside exposure the *human factor*, insisting that people are the most dangerous permanent vector of disease.[3] The authors are completely right when claiming that only a human being can palpate a doe with mastitis and then systematically go on to infect the mammary glands of every other doe palpated that day. But it is important to remember that a family farm is not a hospital, and therefore most don't operate like one. And sterility needn't be the goal at the risk of losing other aspects of nature. (You might want to consider sanitizing your hands in between handling different rabbits, especially when you are coming back into the rabbitry after handling rabbits on pasture.) Our farm is a social space, bustling with CSA members, visitors, volunteers, and our ever-growing team. We also run what we like to call a glass-walls operation, meaning our clients, customers, and community members can see how we grow their food from start to finish. Despite the "biological threats," our farm just isn't a closed space, and no one is putting on paper booties before they step into our stockhouse. While I'm always aware of the fact that circumstances might arise in which we will need to restrict access to our rabbitry, we have thankfully never needed to do so to date.

One place we do exercise extreme caution in our pasture-wire hybrid rabbitry is between our outside tractor rabbits and our inside breeding rabbits. The most common place our rabbits pick up illnesses is through contact with wild rabbits that visit ours while they are out on pasture. For this reason, once we move a rabbit outdoors, we never move it back into our rabbitry. If we need to take an animal off pasture for any reason, we isolate it in a separate barn, so that it has no contact with the rest of our herd. Even perfectly healthy rabbits can be latent carriers of communicable diseases, and it's just not worth the risk.

Protect Your Herd from the Elements

The key to keeping rabbits anywhere outside of the ideal Mediterranean climate, where it's never too hot or too cold and the air is always dry, is keeping your herd protected from the elements. Rabbits are surprisingly adaptive to temperatures as low as −10°F (−23°C) and as high as 95°F (35°C), which is good because in upstate New York, we hit both of those temperatures several times per year. Keeping your rabbits out of the sun, wind, and rain will keep them thriving under most conditions.

A cold rabbit has the ability to regulate its own body temperature in most circumstances. It will instinctively change its body position by curling up to reduce heat loss as well as alter its feed intake to eat more, so that it can expend more calories necessary to keep warm. Studies have even shown that if shaved and then exposed to cold, Californian rabbits will grow black fur in place of their original white fur in order to absorb more heat from the sun.[4] However, these temperature-regulating superpowers become drastically less effective if animals are exposed to wet or windy conditions. So remember, a cold rabbit is fine, but a cold, wet rabbit is in peril. A cold, wet rabbit in the wind doesn't stand a chance.

When it comes to the other side of the weather spectrum, rabbits again have good natural coping instincts. On warm days, rabbits will stretch way out, maximizing the potential for respiration from their big ears and long limbs. Because of all their fur, perspiring is not very effective, so rabbits pant to evacuate heat from their bodies instead. They also naturally restrict their activity until temperatures fall in the evening. However, they cannot manage all this without shade and airflow, and the onus is on the farmer to provide these things if they are not naturally present.

If you don't have a state-of-the-art temperature-controlled barn, don't worry—neither do we. Thankfully, there are other ways to keep your rabbits comfortable during a heat wave. As I have mentioned, our stockhouse is a regular unheated hoop house—the same size and style as the other three we use for vegetable production—and lacks heating or cooling. The features that make our house hospitable for animals are twofold: First, it's wrapped with white polyurethane, which blocks substantially more sunlight from passing through than the clear material we use for the vegetable houses. Second, in spring, summer, and fall, the entire house is covered with a shade cloth, a commercially available stretch of fabric that is normally draped over greenhouses to ensure they remain cool during summer. They're usually made from woven polyester but can also be made from aluminum. Different styles of shade cloth block different amounts of light, the specifics of which are denoted using a percentage. Right now our house has a 60 percent shade cloth, meaning it blocks 60 percent of the UV rays that hit it. This feature, combined with the white poly skin and some heavy-duty fans, has been adequate. We are, however, investing in some 80 or 90 percent cloth for next season, so we will be better prepared for the increasingly common heat waves.

Some summer days in New York are still just too hot for the shade alone to keep our rabbits cool. One year we had almost a full week of 100°F-plus (38°C-plus) weather. It had me pretty nervous, but we moved quickly to roll out some painter's canvas across the top of all of our cages and soak it down with water every couple of hours. The canvas absorbed most of the water, so the rabbits stayed dry, and the air underneath cooled as evaporation occurred.

A simple water fog system does the same thing—but better. When hooked up to a regular garden hose, fogger lines emit a mist of water droplets smaller than the diameter of human hair. When these droplets hit the surrounding air, they flash-evaporate, reducing the surrounding air temperature as much as 20°F (11°C). These systems are easy to install and pretty inexpensive. Complete kits are available from KW Cages and cost roughly $4 per doe if your rabbitry follows specs similar to ours. We recently invested in one of these systems and debuted it on the first 90°F (32°C) day of the year. It worked great for the rabbits, and I won't lie, I enjoyed standing under it, too.

Depending on the layout of your farm, keeping your rabbits cool in extreme weather out on pasture can be a little trickier. If your tractors are near a water source, you can simply set up a misting system right inside. If they are way out in the back fields, like ours sometimes are, your best option is to freeze some 2-liter or gallon-sized water containers and set them inside the tractors for the rabbits to cozy up to and cool down. Keep two sets, so you can swap them out as they melt. I do this only when the weather spikes above 95°F (35°C), and to date we've never lost a rabbit to heat stress out on pasture.

Look for Signs of Heat Stress

While we've never lost a rabbit to heat stress on pasture, unfortunately we have inside the stockhouse. It was during an early spring when we had not yet built our new stockhouse, so our rabbits spent the winter in one of our vegetable greenhouses. Because it was still cool outside, I had not yet moved our rabbits back into the shady old barn in which they spent their summer. For the same reason, no one on our vegetable team had put any shade cloth up on the greenhouse.

That day, the sun finally came out after weeks of rain and overcast. I was in the field working, not thinking about anything in particular, except maybe about how nice it was to see the sun again, when one of our crew members came running to tell me something was wrong with the rabbits. By the time I got there, one of our does had already died. We lost another two does before I even had the chance to figure out what was going on. Then it hit me like a ton of bricks. All of our black-furred rabbits were in peril. Even though it wasn't very warm, the direct sunlight was causing them severe heat stress, and in the worst cases heatstroke. (I still consider that day to be one of my very worst as a farmer. But my hope is that after reading my account, you will not accidentally do something similar. I also want you to know that even so-called experts make very big mistakes from time to time—although I can assure you that I have never, nor will I ever, make this particular one again.)

As the name implies, heatstroke is the sudden rise in body temperature. At its peak, this couples with neurological and physical symptoms stemming from the animal's inability to dissipate heat. When you aren't panicking like I was, the signs of heat stress are easy to recognize. Look for lethargy, unresponsiveness, excessive panting (the key word here being *excessive*, as regular old panting is normal for rabbits on warm days), and reddening of the skin around their ears. As the stress worsens, rabbits will begin to drool in a last-ditch effort to cool off. Drooling is not a normal behavior for any type of rabbit, so if you see moisture around the face and mouth it's very likely they are experiencing heatstroke, and you should move quickly to cool down the afflicted animals. Failing to act fast can lead to convulsions, followed by coma and imminent death.

To cool down a rabbit quickly without sending it into shock, first get it into the shade or into an air-conditioned space. Next either mist the rabbit with a spray bottle, wrap it in a damp towel, or place it in a shallow vessel filled with water. Always use cool water—not ice cold. Stay calm and handle your animals gently to prevent any additional stress.

I know this all sounds scary, but don't let my one bad experience scare you away from rabbits. All it takes to prevent heatstroke is awareness and attention. Don't get caught without them, and you and your bunnies will be just fine.

Prioritize Air Quality

Even the cleanest, shadiest cages will not compensate for an unsuitable environment. In order to thrive, rabbits need adequate space, the specifications of which were detailed in chapter 3. They also need good air quality. And while maintaining healthy air is easy for your rabbits on pasture—there is plenty of ventilation outdoors—it can be tricky in barns or other closed facilities. Ammonia levels in the indoor air should be the primary concern, because too much ammonia can severely weaken a rabbit's upper respiratory tract, and a degraded respiratory system opens the door up for bacterial infections.

Caused by the buildup of noxious gas from naturally degrading urine, ammonia is measured in parts per million (ppm), and for livestock 20 ppm is the threshold at which levels are considered high. Even so, it's recommended that producers work to keep ammonia levels below 10 ppm.[5] This is easy for us in the summer, because our breeding rabbits are housed in a 96 × 30-foot (29.3 × 9.1 m) hoop house that gets plenty of ventilation. In the warmer months, all we need to do to keep ammonia levels low is open up the doors, roll up the sides, and let the fresh air pour in. Plus, in the summer we have only our rabbits and our baby chicks in the stockhouse, so the overall production of ammonia is fairly low.

In the winter, however, it's a different story for two reasons. For one thing, it's cold, so we keep the doors closed and the sides rolled down to keep everyone warm and to keep the water lines thawed. Second, we have a lot more animals in the hoop house during the winter when the pasture is inhospitable due to the cold, wind, and wet. During spring, summer, and fall our hoop house is for breeding rabbits and our baby animals, but in the winter it also hosts our flock of laying hens, our breeding rabbits and any litters born after September, and five or six feeder pigs. Depending on the weather, we may even have a batch or two of early chicks brooding in there for a few weeks. I wouldn't use the word *crowded*, because we work very hard not to overstock our animals, but the hoop house is definitely used to its fullest capacity in January, February, and March. These are the months when we need to get serious about monitoring our air quality.

Healthy barns and stockhouses don't really smell like anything except dirt, hay, and maybe wood shavings. If a non-farmer walks into your rabbitry and that's all they smell, your air quality is likely just fine. But if a farmer walks into their rabbitry and that's all they smell, there could still be a problem. This is because the average person can smell ammonia pretty clearly, even at just 20 ppm. The problem for farmers, though, is that regular exposure desensitizes us to the point that we just don't notice it anymore, even if the levels are high. To remedy this, I use ammonia test strips to check our levels. They are

We keep the sides rolled up and the doors open whenever we can in order to increase airflow in the stockhouse.

cheap, quick to use, and accurate enough for our purposes. You can buy these test strips online. To use them, simply tear off a 1-inch (2.5 cm) strip and wet it with a couple of drops of clean, distilled water. The paper will change color in a few seconds and can be compared to the accompanying chart to tell you how much ammonia is in the air.

Our stockhouse has highly permeable dirt floors and good drainage, so when ammonia levels are high it's not usually because of our rabbits, as their urine just seeps into the ground and drains away. It's usually our laying hens who are causing the issue, because their bedding either isn't dry enough, isn't deep enough, isn't clean enough, or any combination of those things. If ammonia levels are high in your rabbitry, you should:

Increase ventilation to whatever degree you can without exposing your animals to any extreme weather.

Remove manure more regularly.

Add an absorbent substrate underneath your cages to absorb the urine between cleanings. (Note that pine shavings are better than hay and straw, which are not very absorbent and tend to matt up, making them difficult to remove.)

If you are doing all three of these things and ammonia levels are still too high, your facility may too densely stocked. There is also a connection between high ammonia levels and diet. If there is too much protein in your rabbits' diet, the excess will be excreted via their urine as nitrogen. This, when mixed with the bacteria present in your rabbitry, will turn to ammonia. Look for flies when determining if diet is the cause of ammonia in your barn, as they are also attracted to the extra nitrogen and can serve as a good indicator of this issue.[6]

Rotate Animals Through Fresh Pasture

Raising rabbits in a pasture-based system adds another element to practicing good preventive hygiene that conventional rabbitries don't have to worry about. Rabbits are susceptible to all the major parasite families found on pasture including flukes, tapeworms, intestinal worms, and coccidia, among others. Many of these can survive in soil for quite some time, even without a host present. Because rabbits are sensitive to parasites, it is important to keep them on fresh grass in order to limit their exposure. A report from the Maine Organic Farmers and Gardeners Association recommends rabbits not revisit the same ground for an entire year, but in our experience three months is enough time.[7] However, if you encounter a decline in your rabbits' health or performance, you should definitely increase the rest time for your pastures.

For the most part, following these guidelines will keep your rabbitry running smoothly. But at the end of the day, disease is all around us, and no pasture-based livestock operation remains untouched forever. When something erupts, stay calm and try not to get wrapped up in what the internet says. It's full of pet owners and amateur homesteaders—not livestock veterinarians and experienced production farmers. With the rare exception, most of the advice found there will not be very helpful, and a lot of what you read will just make you feel stressed. If you can find one, build a relationship with a small livestock veterinarian, and get in touch with other rabbit producers near you. They will be your best resources in times of trouble. If you need help getting connected to the right people, reach out to your local Cooperative Extension agent—it's their job to help farmers access information, and in my experience they're really good at it!

Common Ailments

As with all living things, there is a long list of things that could potentially go wrong with your rabbits. The pages that follow are by no means comprehensive, but they cover the more common as well as the most serious ailments.

As a source of information, I like to use the MediRabbit website when I need help diagnosing an issue.[8] MediRabbit is an educative nonprofit based in Geneva, Switzerland run by Dr. Esther van Praag, a rabbit biology and disease specialist. In the past their staff has kindly responded to my inquiries and offered well-researched and helpful advice. I always try to give a donation in return.

BASIC PROTOCOL FOR SICK RABBITS

No matter what the issue is, there are six basic steps you should take upon encountering a sick rabbit in your herd. The following list is adapted from a publication from the Pacific Northwest Extension Office:[9]

1. Note the cages, pens, or tractors that contain sick rabbits.
2. Isolate the sick rabbits from the rest of the herd. Ideally, move them to a separate room or building, but if this isn't possible, move them as far from your healthy animals as you can and make sure there is no direct contact.
3. Disinfect your hands before handling your healthy rabbits again. Moving forward, care first for the healthy animals, moving on to your sick animals last to prevent the spread of disease via your hands or tools. Disinfect your tools and hands (boots as well, if you can) before returning to your healthy animals.
4. Diagnose the problem and begin the appropriate course of treatment. If you can't diagnose the issue on your own, contact a specialist.
5. Cull all hopelessly sick animals and bury them away from your rabbitry and pasture.
6. Clean and disinfect all cages, pens, and tractors that contained sick animals before placing new rabbits in them.

Abscesses

Abscesses in rabbits are pretty much the same thing as they are in humans. When diagnosing one, look for a swollen area on the body containing an accumulation of pus. In rabbits, they are especially common under the skin near the jaw or wherever there is a wound or a scratch. Abscesses are bacterial infections for which there are many potential causes, but most often they are caused by dental disease or injury.

If you have ever had an abscess yourself, you probably already know your options for treatment: antibiotics and lancing. Unfortunately, oral antibiotics are not usually very effective when it comes to rabbits and abscesses, so if you want to treat an abscess, you'll need to go with plan B and lance. Lancing means opening up the abscess and removing the pus inside. While farmers with a steady hand and a strong stomach can do this themselves, typically draining an abscess is best left to veterinarians. Once the pus has been drained, the wound needs to be disinfected with peroxide and packed with an antibiotic ointment in order to heal.

Because abscesses can be pervasive, difficult and costly to treat, and a potential indicator of a rabbit with a weak immune system, we would most likely choose to cull an infected rabbit rather than treat it.

Blocked Mammary Gland

Blocked mammary glands are another thing you may have experienced yourself at some point. As with humans, sometimes nursing rabbit moms will produce more milk than their babies are able to drink. This can cause a clog in the duct.

Look for swollen, hot, and firm mammary gland(s) as your sign that something is amiss—but be careful not to mistake a simple blocked mammary gland with *mastitis*, a bacterial infection (that I will discuss later in this chapter) that can cause similar symptoms. To remedy a clogged duct, massage the gland with a warm cloth, and reduce the buildup of milk with partial milking. Make sure not to drain the milk completely, though, as that will stimulate more milk production.

Enteritis (Diarrhea)

Enteritis is gastrointestinal disease caused by inflammation in a rabbit's intestines. It's pretty easy to identify—just look at the poop. If it is soft, loose, mucus-y, or watery, there is definitely some intestinal drama going on. Other signs include messy, stool-covered tushes, weight loss, dehydration, ruffled-up-looking fur, and general poor body condition. As you spend more time with rabbits, you will simply know it when you see it. While this might sound crazy, I can actually smell an enteritis issue before there are any visible symptoms.

While enteritis is easy to diagnose, it is harder to identify the cause. Basically, any factor that alters the bacteria in the cecum, of which there are many, can be the root of this problem. Unlike the case of humans, where diarrhea just happens sometimes and it's no big deal, intestinal problems in rabbits are unfortunately often fatal. Improper diet, such as one too low in fiber or too rich in carbohydrates or sugar, or sudden change in diet, stress, and environmental issues can all cause diarrhea. In cases like these, changing the feed and improving the environmental conditions can clear up the problem.

In other cases, enteritis can be caused by viruses or pathogenic bacterial growth, like *Pasteurella multocida*, which I will talk about later in this chapter. Intestinal parasites (like tapeworms) and *protozoans* (coccidiosis) also cause diarrhea. In each of these cases, the enteritis is just a symptom of a larger issue, and the treatment will depend on which larger issue it is. Unfortunately, it is often true that by the time a rabbit has diarrhea, the situation has gotten too severe for treatment. For this reason, we typically cull or slaughter our rabbits at the first sign of enteritis.

Hairballs

Like cats, dogs, and other furry animals, rabbits can get hairballs—a collection of fur inside the stomach. Most common in long-haired breeds like Angoras, they are caused by rapid and excessive ingestion of fur. In mild cases this will cause your rabbit's fecal pellets—which are typically little individual balls—to be strung together. Some farmers call it "a string of pearls," which is kind of gross, but also a fairly accurate description of what it looks like. Rabbits with more severe hairballs will demonstrate a reduced appetite and can have a firm mass in the stomach that can be felt with palpation.

As always, prevention is best. If you make sure there is enough fiber in their diet, brush your rabbits while they are molting, and remove loose fur from their cages, you should rarely experience this issue. If you do experience a nasty hairball, providing an oral dose of approximately ½ ounce (14.8 ml) of mineral oil can help the rabbit pass the fur.

Hutch Burn

Also known as urine burn, hutch burn is caused by regular contact with urine, most often via solid-floor hutches that are not cleaned properly, or dirty resting boards or nesting boxes. It is easy to identify—look for bald spots and chapped, red, inflamed skin. It is particularly present on the legs and around the genital region. The result of environmental factors, hutch burn is not infectious or contagious. All you need to do to prevent it is improve your rabbit housing to ensure it stays clean and dry. To treat a case of hutch burn, gently rinse the affected rabbits with warm water, then pat them dry before returning them to freshly cleaned and dried quarters.

Pregnancy Toxemia or Ketosis

Pregnancy toxemia can cause sluggishness, dull eyes, loss of appetite, and death, which occurs just before or after kindling. It is most common in does pregnant with their first litters, and on our farm we have experienced it during the last week of gestation. While the exact cause is unknown, it is thought to be starvation, which is ironic because this disease disproportionately affects obese rabbits. If the doe does not succumb to the illness before going into labor, usually the delivery of the kits is enough to right the issue. Otherwise it's not really a treatable disease. Your best move against ketosis it to reduce the risk by keeping your does inside their target weight range.

Mastitis

Symptoms of mastitis are similar to those present when a nursing doe has a blocked duct: Mammary glands and teats appear red and swollen and are hot to the touch. One surefire way to tell one ailment from the other is to look for the tissue to appear blue, which it does in some cases of mastitis, but not when there's just a blocked duct. This is why the colloquial term for mastitis is *blue breast*. This condition can be painful, so afflicted does may refuse to nurse their kits. They may also demonstrate reduced appetite.

Mastitis is caused by the presence of bacteria in the mammary glands, most commonly *Staphylococcus*, and antibiotics are necessary to treat it. Therefore, prevention is paramount. Make sure nesting boxes have smooth edges to prevent trauma to the teats, and sanitize boxes between litters. Do not move kits from the infected doe to a foster doe, because they can spread the infection. Also, be sure to disinfect your hands before handling other rabbits to avoid becoming an unintentional vector.

Rabbit Syphilis or Vent Disease

Like people, rabbits can get venereal diseases, and this is one of them. You can identify a syphilis outbreak by watching for blisters, scabs, and inflammation of the genitals. Red sores on the mouth and face can also be an indicator. This condition is caused by the spirochete *Treponema paraluiscuniculi*, and it is found in both sexes.

Breeding does can pass the disease on to their offspring, so you will want to get an outbreak under control quickly. To do so, check with your veterinarian to get the proper antibiotic, and treat all any rabbits that have been in contact with one another, even if they are not showing symptoms. Not all carriers will have an outbreak at the same time, so this is key to making sure you get rid of the disease for good. Once the syphilis has been eradicated and the rabbits have gone through the required withdrawal time, breeding can continue as usual.

Sore Hocks/Bumblefoot

Sore hocks are abscesses or calluses on the underside of a rabbit's legs and on the feet and foot pads. They are often the result of stress on a rabbit's joints or feet and can be caused by improper support in wire cages, obesity, and unclean or damp housing. Any break in the skin provides an opportunity for the area to get infected, resulting in an abscess.

To avoid sore hocks in your herd, make sure your housing is clean and dry, and that the flooring has enough support, so that it does not cause unnatural bowing of your rabbits' legs. It is best to add a wooden plank for your rabbit to rest on while the condition heals. If the problem is pervasive, there is likely an issue with your housing setup, and you should contact a veterinarian or an extension agent for a site visit to help identify the cause.

Ear Mites

If you notice your rabbits are excessively scratching their ears or shaking their heads, you may have ear mites in your barn. Look for waxy brown buildup in the ears or any visible scabs, as they are good indicators of this problem. Ear mites are teeny-tiny parasitic arachnids that can show up out of nowhere. They are especially common in rabbits housed outdoors and in tractors, but if you have them in your barn they may have hitchhiked on some straw or hay and spread from there.

Though icky, ear mites are treatable. To get rid of them, start by separating any infected rabbits from the rest of the herd. Next, disinfect all housing, waterers, feeders, and nesting boxes with a diluted bleach solution. In mild cases adding a few drops of vegetable oil into the affected ears will smother the mites and eliminate the problem. You will need to repeat this process every other day for a couple of weeks and then twice more over the following fortnight to keep the mites at bay. Do not remove any scabs, as they will fall off on their own. Once treatment has been completed, the rabbit can be returned to the herd.

If the infection is severe, you can treat breeding rabbits with ivermectin, a medication that is used to treat a variety of parasites.[10] I have never seen an ivermectin labeled for rabbits, so check with your vet to confirm the proper dosage and length of treatment. It's pretty strong stuff, and it has a fairly long withdrawal time of two to four weeks.

Coccidiosis

Anyone who has been raising animals for a few years will likely be able to tell you about an experience they've had with the dreaded coccidiosis—and you've already heard one of mine. It's a common sporozoal infection that can afflict virtually any type of livestock. Known as cocci for short, this prolific parasitic disease is thankfully host-specific, so an outbreak in your chickens will not affect your sheep, for example.

The disease-causing parasites, called Coccidia, are often just present in the environment, but in our experience the outbreaks become particularly prevalent during warm, damp, and humid weather. After they are ingested by an animal, they proliferate in the gut and spread to other hosts through fecal matter. This process of spreading is known as *fecal-oral transmission*. Once Coccidia are present in the environment, they are very difficult to get rid of—oocysts can survive and remain viable without a host for over a year. In rabbits, cocci infections cause damage to the liver and/or the intestines, which over time can result in serious illness or death.

To identify a cocci outbreak, look for soft stool, diarrhea, weight loss, slower growth, and loss of appetite. The sudden onset of poor body condition is a good indicator. There are two types of infections: *hepatic* and *enteric*. If the infection is hepatic, at necropsy you will find a telltale enlarged liver with many white pockmarks all over.[11] If it's enteric, like our outbreak this winter was, you'll notice a gas-filled and bloated GI tract.

This issue can be treated with coccidiostats, like amprolium (we typically use Corid), and antimicrobials, like sulfadimidine, which can be purchased through online veterinary pharmacies and at some farm stores, like Tractor Supply.[12] However, we generally choose early slaughter whenever we can, unless the issue is in our breeding stock or we catch it very early in our juveniles. The key to ending the cycle is practicing good sanitation. Make sure cages and nesting boxes are clean and poop-free; move your rabbit tractors at least once every day, and more if you experience an outbreak. Give twigs that are high in tannins, like those from willows and pear trees, as a preventive measure.

Snuffles/Pasteurella

Runny eyes, runny nose, and frequent sneezing are the telltale signs of the snuffles (which is the technical term, I promise). As the illness advances, infected rabbits may become lethargic and stop eating or drinking, resulting in weight loss and dehydration. There are a number of other health problems that can result from snuffles, including neurological disorders and abscesses.

This disease is the result of a bacterial infection, usually caused by *Pasteurella multocida*, which is why it is sometimes known as pasteurella. Rabbits with strong immune systems can carry loads of this bacteria without issue, but weak or stressed animals may experience severe health issues as a result. Pasteurella can be picked up almost anywhere, although animals in poorly ventilated, dusty, moldy, or damp settings are particularly vulnerable. Likewise, animals kept on pasture are more at risk, as the bacteria can be in the soil or carried by wild rabbits who could have contact with your herd.

Snuffles is a tricky disease to treat, especially if you are trying to avoid antibiotics. If you see signs of an infected rabbit, quarantine it immediately, as the disease can spread quickly to vulnerable animals. You should also sanitize all the feeders, waterers, and housing with a diluted bleach mixture. If the sick animal was out on pasture, move your tractors to fresh ground to reduce potential spreading. Some infected animals will recover on their own, while others will continue to worsen. Animals who fail to recover should be culled to reduce potential spreading and to end suffering. Snuffles can be effectively treated with antibiotics obtained with a veterinarian's prescriptions, but you should act quickly if taking this approach.

Rabbits' susceptibility to pasteurella is a major reason they are so rarely raised outdoors in commercial production. The bacteria is so prevalent in the natural world that it is almost impossible to avoid it forever. We experienced our first outbreak after two full seasons without issue, and it scared the heck out of us. Rabbits who were perfectly healthy one day were suddenly lethargic and emaciated the next. After a very expensive trip to the vet and a $200 necropsy, we were told that all we could do was clean the barn, sanitize all the equipment, move the tractors to fresh ground, and let nature take its course. And that it did. We lost more than a dozen young rabbits in just a week, and we were terrified we would soon lose the entire herd. Thankfully, that never happened (despite what the internet said). The strongest rabbits either recovered or never became sick, and we were left with those who had a natural resistance to the disease. A prominent rabbit producer I know told me he experienced the same thing when he first put his rabbits on pasture. The only way to truly guard against the snuffles is to breed for resistance, and unfortunately, it takes some time.

Conjunctivitis

If your rabbits have red, swollen eyes with fluid leaking out, or their eyelids are stuck shut, they are probably suffering from conjunctivitis. This is a bacterial infection in rabbits that is similar to pink eye in humans. It can be the result of overcrowding, injury, or unsanitary housing. The most common bacteria involved are *Staphylococcus* and *Pasteurella*.[13]

To treat conjunctivitis, open up the eyelids (if they are stuck shut) and clean the surrounding tissue with a warm cloth. Flush the eye with a boric acid or sterile saline solution (the same kind you would use for human eyes) and apply a thin coat of a general-purpose antibiotic ointment, which can be found at most farm supply stores. It should clear up in a few days.

Wry Neck

A rabbit may be experiencing wry neck if it constantly leans its head to the side or has a head that is stuck in a crooked, inquisitive position. Severely afflicted animals may experience dizziness resulting in rolling and difficulty eating. This is a condition with many possible causes, including ear infection, neurological issues, parasites, head injury, brain tumor, and stroke.

The course of treatment depends on the specific cause, making this one a little tricky. If the head tilt is only slight, it is probably the result of an ear infection. The only organic treatment for this is using a saline wash in the ear, but a topical antibiotic will work as well. Otherwise severe head tilt is very hard to cure, so to prevent suffering either cull or butcher the animal.

Myxomatosis

Myxomatosis (or colloquially, *myxo*) is caused by the Sanarelli virus, which wild rabbits can carry without experiencing any effects. This virus devastated the rabbit populations all over Europe in the 1960s and 1970s due to its contagious nature as well as how easily it can be spread over wide distances via biting insects, like mosquitoes and fleas.[14] This is how it managed to spread so rapidly across an entire continent. It can also be spread via rabbit-to-rabbit contact and shared equipment. Certain strains of myxo can cause widespread death within one week. Thankfully, myxo outbreaks are exceedingly rare these days, and less than 20 cases have been reported domestically over the past decade—every one of which was in California or Oregon.[15]

There is no treatment for myxomatosis, and infected animals should be promptly culled. Vaccines for the disease are used in Asia and Europe, but they are not currently available in the United States.

Myxomatosis is a reportable disease, and any outbreaks should be expeditiously shared with the USDA.[16]

Fur Mites

You'll know your rabbits have fur mites if you witness frequent scratching, loss of fur, and lots of dandruff. These symptoms often present at the base of the ears and between the shoulder blades and are the result of an infestation of non-burrowing skin mites. You can confirm their presence by looking very, very carefully for the little buggers. They are tiny, but not microscopic.

Healthy rabbits can host small amounts of these mites without experiencing any symptoms, while weak rabbits are more greatly impacted. Ivermectin is an effective treatment, but again I would use it only on breeding stock since the withdrawal time for rabbits for consumption is two weeks, or four if you are doubling it as the Animal Welfare Approved board recommends.[17] To prevent further infestation, clean any areas that potentially harbor mites such as nest boxes or wooden hutches. Make sure hay and straw are fresh and clean and that other animals in your barn, like dogs and cats, are also free of mites. Quarantine new rabbits for at least two weeks before integrating the rest of their herd.

Warbles/Botfly Infestation

If you see a lump under the skin that looks kind of like a tumor, but with a distinctive round hole in the center, your rabbit may have warbles. Warbles is a nightmarish ailment caused when a botfly deposits eggs into a rabbit's skin. As the fly larvae mature, a large, hard mass grows under the skin. Eventually they emerge from their host's skin as fat grubs. The distinctive hole is actually the breathing hole for the larvae; squeezing around it gently may reveal a protruding larva.

Warbles is treatable, but in the spirit of full disclosure, I'll let you know I've never had a botfly infestation and therefore have thankfully never needed to treat one. I do know, however, that the key is to cut off the air supply for the grubs inside, so they'll be forced to crawl out of your rabbit in search of oxygen. To do this, coat that little breathing hole with a generous schmear of Vaseline and wait. When the creepy crawlies start making their way out, extract them very carefully using a pair of tweezers, but be careful not to kill them. You need to get the entire grub out, as left-behind parts could cause a secondary infection. When you are done, clean the wound well with germicidal soap and give yourself a round of applause—you're officially the bravest person in the world.[18]

Malocclusion/Buck Teeth

Malocclusion is just a fancy word for overgrown teeth. It's an inheritable condition that occurs when the lower jaw is shorter or longer than the upper jaw, or if the teeth are damaged in some way. Malocclusion can be corrected temporarily by trimming the teeth, so the animals can eat and grow before slaughter, but any adult rabbits with this condition should not be bred, as the condition is hereditary. The process of teeth trimming is not painful for rabbits so long as it's done right. It's similar to trimming nails and is typically done using regular pair of wire cutters. However, I recommend having a trained veterinarian show you how to do it the first time, as incorrect trimming can cause the teeth to split or crack.

What to Do When You Don't Know What to Do

There may come a time when you just can't diagnose what's wrong with your animals. If this happens, there are a few things you can do. The first is to call your veterinarian. Finding a good vet for your rabbitry might take a little time, so be sure to do your research before you start to build your herd. The reason finding one could be a challenge is because the majority who work with livestock are large animal veterinarians, which means they specialize in bigger animals like cattle, horses, sheep, and others. Small animal vets usually work with rabbits, but this is generally in the context of them as pets and not livestock. Veterinarians who work exclusively with pet owners might suggest courses of treatment that simply don't make sense on the farm, because they either are too costly or require the use of medicine not suitable in meat production.

Another option is to contact an extension agent or specialist. Extension agents are university employees who develop and deliver educational programs to assist people in economic and community development, leadership, family issues, agriculture, and environment. Extension specialists are subject matter experts in fields ranging from agriculture to life sciences, economics, engineering, food safety, pest management, veterinary medicine, and various other allied disciplines. Both agents and specialists work within the larger Cooperative Extension System, and it's their job to help you be good at your job. I never hesitate to call the livestock expert at Cornell Cooperative Extension when I need some advice.

If your research, your vet, and your extension specialist can't seem to help you, you can bring a deceased rabbit to a diagnostic laboratory in order to learn exactly what the cause of death was. It's easiest to drive your sample if the lab is close by. Otherwise there are specific regulations for shipping frozen carcasses of diseased animals through the mail. When I had to do it, I asked my veterinarian to arrange it for me; she knew the proper protocol and was happy to oblige. If you are going to arrange shipment yourself, be sure to check with the US Postal Service and the lab to which you are shipping beforehand to ensure that your specimen arrives at the right time and in the correct condition.

The more information you can provide the pathologist, the more useful they can be. As described in a guide from the Oregon State Cooperative Extension called *Domestic Rabbit Diseases and Parasites*, it's helpful to include a letter detailing the following when sending a specimen:

Number of rabbits on the farm.
Number of sick or dead rabbits.
Age and sex of affected rabbits.
Description of the disease as you observed it—for example, "Rabbits develop watery diarrhea, quit eating and drinking, and die in 1 or 2 days."
Dates of first losses and subsequent losses.
Incidence of infection (whether it is in just one house or pen, or scattered throughout the rabbitry).

NORMAL RABBIT VITAL SIGNS

Per the Merck Veterinary Manual, normal rabbit vital signs are: [19]

Body temperature: 101.5–103°F (38.6–39.4°C).
Rectal temperature: 103.3–104°F (39.6–40°C).
Heart rate (pulse): 130–325 beats per minute.
Respiratory rate: 32–60 breaths per minute.

What treatment, if any, has been given.
Type and brand of feed and feeding methods used for the past six months.
Type of housing (whether the rabbits are kept on wire, pasture, or solid floors).
Any other information that might help explain the outbreak.[20]

The American Association of Veterinary Laboratory Diagnosticians has a comprehensive list of their labs organized by state available on their website.[21] Be aware, it is pricey to have specimens diagnosed. When we did so, it cost around $200 per animal. In our case, we had an outbreak that seemed to have the potential to be severe, so I thought it was important to know exactly what was going on. You should not feel compelled to send every single dead rabbit you encounter to a diagnostics lab. Use your discretion.

It only took about a week to receive the results from our testing. The report was very detailed, but also super dense. I took it over to our vet, who helped me understand and analyze the information. She did this pro bono, as I assume most would, but if not, this is another opportunity to get in touch with your local extension specialist or to talk to the agents at the laboratory who did the testing.

When to Cull a Rabbit

Some people use the terms *cull*, *slaughter*, and *butcher* interchangeably. On our farm, they each refer to different actions. Slaughtering (or *processing*, which is what we usually say on our farm because it's easier for regular civilians to swallow) is the term we use anytime we are killing animals for food. If we are crating up chickens or rabbits for slaughter, they are regular, healthy animals that have reached our prerequisite weights for selling. When we *cull* an animal, it means we are choosing to euthanize (kill, but not for food) or slaughter it early, as a means of either putting an end to its suffering or preventing an illness from progressing or spreading. We also use the word *cull* when slaughtering an older animal that is no longer productive or a younger one with undesirable traits we don't want to risk passing down by breeding. The word *butcher* refers to the breaking down of a processed animal into retail cuts.

The question you will undoubtedly need to answer at some point, whether you are raising rabbits commercially or in your backyard, is whether to treat a sick animal or to cull it. Every farmer I know approaches this tough question differently, and unfortunately there are no hard-and-fast rules. When trying to decide how to write about culling rabbits without sounding like a monster, I looked to what some other livestock farmers have said about the subject. As described in his book *The Small-Scale Poultry Flock*, Harvey Ussery uses a logic similar to mine with his chickens:

> *If you want to treat a sick chicken and try to nurse it back to health, I wish you success. Certainly, if you have a flock of half a dozen hens, the loss of one will be a proportionately greater loss than in my flock of dozens. But the theme of this book is husbandry of the productive small-scale flock, and therefore using poultry as a serious partner for food independence; that is, management decisions look to the future as well as the present. . . . Seen in that perspective, my decision to cull a diseased bird from the flock immediately is hardly as callous as the reader may first assume. My concern is*

not only to provide as healthful and enjoyable a life as I can for today's flock but to help ensure the well-being of all its future members as well. Especially for someone who breeds some of his own stock as I do, is it kindness or cruelty to rescue a sick bird and thus pass her genes on to future progeny?[22]

Honestly, Harvey, I couldn't agree more.

————————

My reasoning for including a list of potential ailments and illnesses in this chapter is not to encourage you to treat every single issue that arises. You may have been able to tell, as my descriptions of how to treat most of the illnesses are fairly vague. At our farm we rarely treat any of our animals with antibiotics. In general, we do not use medication with our small livestock because they simply are not alive very long before they are processed for meat, and even though we could hold them until after they meet the required withdrawal times, we prefer culling. This is not to say I would *never* give an antibiotic or another kind of medicine to our animals. If early culling is not a good option—for example, if the rabbits are still very young and small, or if the issue is in our breeding stock—I will choose to pursue a course of treatment.

When it comes to rabbits, generous culling can be the key to building up a healthy, resilient herd. If a rabbit is demonstrating early symptoms but can still be comfortably and safely transported to our slaughterhouse, it is usually our choice to do so. For many ailments, the risk of transmission or the cost of treatment is too high for it to make sense for us to try to nurse the animal back to health. And like Mr. Ussery, we want to pass on only the genetics of our healthiest animals. However, it is important for livestock farmers to be good diagnosticians, even if we are not good doctors or surgeons. And even if you are going to cull your sick or injured animals, you still need to know *what* caused the problem. Being able to identify an issue is the first step in preventing it from happening again.

CHAPTER 10

Processing Your Rabbits

If rabbits happen to be your first foray into raising animals for food, coming to terms with the slaughter will likely be challenging at first. I have been raising livestock for the better part of a decade now, and still I hate watching our pigs walk off the trailer at our slaughterhouse. Taking the life of another living thing is never easy, but it definitely becomes easier once you develop the skills to do it deftly.

I have people ask me all the time at the farmers market if I ever do the slaughtering of our livestock myself. "Yes," I say, because even though we use a slaughterhouse now, we did all of our own processing on the farm for several years and still do on rare occasion. Often this is met with something like a gasp and an "I could never . . ." Sometimes I ignore their shock and awe, and other times I say something like, "Well don't worry—I do it so you don't have to." I know it sounds a little snarky—honestly, that's probably why I say it—but it's also true. Livestock farmers and slaughterhouse technicians wrestle with the moral quandaries of killing animals for food so that the general public doesn't have to. While some folks imply that I should be ashamed at my ability to wring a chicken's neck (yes, literally—this old-school method is still the best way to cull a sick or injured chicken when I need to) and still sleep at night, I consider myself a civil servant. In an omnivorous society, *someone* has to slaughter animals—why shouldn't it be me?

I also believe that a lot of meat-eaters who say they could never be the ones to dispatch an animal possibly feel that way because they don't know how to do it. I remember holding a knife at the neck of the first animal I ever slaughtered (a chicken) for at least half an hour before I could commit to actually using it. I was terrified that I would freeze halfway through the process and accidentally torture the animal. I won't lie—my first slaughter was neither as quick nor as painless as I desperately wanted it to be. It was sloppy and riddled with insecurity that no doubt made the experience a poor one for the animal. But my second one was better, and my third was perfect. Learning the process all over again with rabbits went pretty much the same way.

When to Process Your Rabbits

If all goes according to plan, your litters will be ready to process at 12 to 16 weeks of age. We pull rabbits for slaughter when they have reached between 5.2 and 6.5 pounds (2.4–3 kg), live weight. Our rabbits typically dress out at 55 percent of their live weight, with the head removed, so pulling at this size gives us 2.9- to 3.5-pound (1.3–1.6 kg) rabbits for our markets—perfect fryers. If your market is for roasters or stewers, you will be holding on to your rabbits longer. If this is the case, don't forget to separate the males and females after 16 weeks so they don't begin to breed.

This relationship of the weight of a dressed carcass to the weight of the live animal is called the *dressing percentage*, and it is different for every breed and individual herd. At 55 percent, our dress percentage is average by industry standards. It's good, but not especially so. The better the meat characteristics are in your herd, the higher this percentage will be. With good genetics and careful breeding, it's possible to get as high as the mid-60s. When comparing your percentages with industry standards, include the liver, heart, and kidneys in your final dressed weight.

Eyeballing live weight is a tricky business, so do not hesitate to get out the scale. At $9 per pound, pulling too many underweight rabbits can wreak havoc on our year-end sales numbers, so it's worth taking

We weigh our rabbits using a hanging scale, so they don't hop away.

the extra step while developing your intuition. We use a hanging scale with a box we can put the rabbits in, because getting them to stay still on a regular tabletop scale is almost impossible.

Depending on genetics and your chosen method of raising, your rabbits may reach market weight in as little as eight weeks. This is the gold standard for conventional rabbitries, but alas, at our farm we have never even come close. We are pretty much always behind in efficiency whenever I compare our numbers with the ones they get at

151

the big factory farms, and that's fine by me. I know it comes from giving our animals higher-quality diets, more room to move around, and no antibiotics. What we lose in efficiency we gain in animal welfare, omega-3 fatty acids, texture, and flavor.

Know Your Rules and Regulations

To avoid doing business in the gray area, you will need to find or develop a good and legal method for processing your rabbits. I recommend researching your options well before you get your first breeding stock, since on-farm slaughter is restricted or prohibited in many states.

Unlike chickens, pigs, sheep, and cattle, rabbits are considered *exotic* or *non-amenable* livestock. This means that they are not included in the Federal Meat Inspection Act as species the USDA is funded or required by the government to inspect. In order to compensate for this fact, producers of non-amenable livestock—like ostrich, deer, and pheasants—in any state can opt to pay for what known as *voluntary inspection*. In these cases, producers themselves hire an off-duty USDA inspector to oversee their slaughter and processing. Product that passes voluntary inspection is stamped and free to be sold anywhere in the country, just like the amenable meats. However, this method is costly and generally impractical for small producers. It costs almost $100 per hour to hire the inspector. You can process a lot of rabbits in an hour, so that's not so much the problem. Rather, the issue lies in the fact that the inspector must remain on-site and be paid for the duration of the *entire* process, even the cool-down, butchering, and packaging. Given that the chilling alone takes a few hours, this all starts to add up, so it's really only practical for big companies that are processing huge quantities of animals at large slaughterhouses. D'Artagnan, for example, processes rabbit via voluntary inspection.

But don't worry—many states have their own statewide regulations and inspected facilities that allow farmers to circumvent these federal guidelines. The regulations for processing rabbits vary wildly from state to state, so you will need to contact your local department of agriculture and markets as well as your health department to learn exactly what is permitted in your neck of the woods. Don't be surprised if it's not easy to find out what is legal. In my experience working with various state agencies, most of the people who work there have no idea what the regulations are. You may just need to keep calling until you finally reach the one person who does.

As of time of writing, New York—where we farm—has an independent, state-run inspection program for exotic livestock as documented in New York State Agriculture and Markets Law Article 5-A. Therefore, folks here can use any slaughterhouse with a 5-A license to process their rabbits. These animals can be sold anywhere within state lines, including farmers markets, restaurants, grocery stores, and on-farm. New York producers, beware—rabbits *do not* qualify for New York State's PL 90-492 exemption (more commonly known as the 1,000 bird exemption), which is the law that allows farmers to process up to 1,000 birds on their farm without any license of inspection. Therefore, technically no farm is allowed to process for sale any rabbits on-site unless they obtain a 5-A license.[1]

In my research for this book, I spoke to regulators in several states in an effort to garner a sense of what is most common. I was blown away by just how much the rules can vary state to state. A few states operate similarly to how we do in New York, with state-run inspection programs. Sometimes smaller states without their own state-run inspection programs allow their producers to use legal slaughterhouses in neighboring states that do have state-run inspection programs.

In other states farmers are allowed to process rabbits for sale on their own farm, without submitting to inspection of any kind, so long as this occurs via *pre-sold direct-to-consumer sales*. This means:

Rabbits must be sold live and to the person who intends to use them.
Money must be exchanged before the animal is slaughtered.
The farmer can then choose to slaughter the animal on-site—
essentially as a favor.
Dressed rabbits may not be sold to restaurants, to grocery stores,
or at farmers markets.

If you are planning to raise and sell rabbits this way, be sure to keep clean and accurate sales records, in case you are asked to present them to the health department or department of agriculture.

Some places have producer exemptions for slaughter and inspection of rabbits. Commonly, these exemptions are for farmers who raise 1,000 or fewer rabbits per year, who are permitted to process their own rabbits for sale within their respective state lines. On occasion I came across a state that allows unlimited, uninspected on-site slaughter for rabbit farms, but in general this scenario is rare. The rules for processing and selling meat are often changing, so be sure to contact your own state agencies for up-to-date information.

The Economics of Processing

Our farm is in New York, so we use a licensed 5-A facility for all of our rabbit and poultry processing. Since good slaughterhouses are hard to find around here, we drive 70 miles (113 km) each way to get there. We also pay $5 per rabbit, which adds up fast. Processing makes up 25 percent of our overall expenses for this enterprise, and it would be much higher if the rabbits weren't piggybacking on our meat birds, whom we take to the same processor on the same day in much higher volume.

Before you start your rabbitry, it's important to decide how you are going to process your rabbits and run the numbers. If you live in a state that allows uninspected on-site processing, you may want to consider doing the work yourself. Rabbits are simple to process and package, and they do not require any expensive specialized equipment, like a chicken plucker or a scalder. If you can legally process your own rabbits, you will save a lot of money (more on slaughtering rabbits yourself later in this chapter).

If you are using a slaughterhouse like we do, make sure the numbers work for you as far as fees, travel time, and mileage are concerned. If they don't, see if you can get creative. For example, if your slaughterhouse is too far away for weekly trips, look for accounts that will take frozen product instead of fresh. If another farm in your area uses the same slaughterhouse for poultry, see if it makes sense to take turns bringing each other's animals. Some states have mobile facilities that may even come right to you.

We make it work by combining our slaughter day with an office day. I go to the slaughterhouse to drop off every Wednesday, and then while I wait for the animals to be ready, I use the local library to catch up on emails, records, orders, and whatever else. This, in addition to hitchhiking with our meat birds, allows the rabbits to share some overhead costs, increasing what would otherwise be narrow margins.

We love our slaughterhouse, and it's important that you love yours, too. After all, their work represents *your* business. If it's good, your customers will think *you're* good. If it's sloppy, they'll think *you're* sloppy. So if you are going to use a slaughterhouse, take your time and do your research. Talk to other farmers, get references, find samples of their work, and go and meet the folks who will be responsible for the final step in all of your labor. You want to make sure to find a slaughterhouse that values animal welfare as much as you do.

Once you find the right people, treat them well. Book your appointments in advance, show up on time, and try as hard as you can not to cancel. Bring them coffee once in a while, if you can. Running a slaughterhouse is a tough job that not many people want. We have to support them every way we can. After all, without them we livestock farmers can't do our job.

How to Process a Rabbit

As with anything else in farming, there are many different ways to slaughter and dress a rabbit. The photo gallery on pages 156 through 159 shows the method we use at Letterbox Farm.

Grading Your Meat

Grading meat in the United States predominantly exists for beef, pork, and poultry sold by distributors and not so much for small, direct-to-consumer farms. USDA grading guidelines for rabbit do exist, however, and even though it's very unlikely it will ever make sense for you to have your meat graded, knowing the standards is useful. Otherwise, how will you know how your rabbits measure up?

continued on page 159

1. With the rabbit on the ground, gently place a thin sturdy bar, like a broomstick (or in our case, a slightly bowed piece of rebar) just behind the skull. 2. Lightly stand on your bar and ready the rabbit by grabbing hold of the hind legs. In one swift movement step down on the bar and firmly pull the legs straight up in order to separate the vertebrae in the neck, instantly dispatching the animal. Be aware the neurons are still firing, so even though the rabbit is dead, it will continue to move for a few seconds. 3. String the rabbit up, securing it by its feet. 4. Using a sharp knife, slit the throat and let the blood drain. It's easiest to tilt the head backward and pull the skin taut to do so. Getting to this step quickly will prevent the blood from coagulating.

5. Using a very sharp, small knife—like a paring knife—score the skin around the hind leg joint. Gently free the hide from the muscle. Be careful at this stage, because pulling too vigorously can tear the meat from the bone. 6. Working your way to the vent region, separate the skin from the meat using your fingers. Using your knife, cut the hide to form a tube. Be sure to cut around the anus, leaving the skin and fur there for now. 7. Now you can easily pull the hide off the rabbit in one piece. Just gently tug downward. Once you get to the head, you will notice the front legs poke out. Free the forelegs from the hide, stopping at the foot. Use a pair of snips to cut the bone at the foot joint. Cut off the head to release it and the hide from the carcass. 8. Taking care not to slice into the organ connected to the anus, sever the bone between the hind legs, shown here at the knife tip. Twist the organ and anus, and move them out of the way.

9. Cut into the belly just below the genitals, and slice downward past the ribs.
10. Gently pull out the innards. 11. Take care not to rupture the bladder (seen full in this photo) or the bile sac, which is the green tube attached to the liver.
12. Snip off the hind feet.

13. Now that the rabbit is slaughtered and clean, let it cool down in a tub of ice water. Adding a generous amount of salt to the water, while not necessary, will help to draw out any blood that remains.

Graded rabbit carcasses are given either an A, B, or C designation. Just as with beef, pork, and chicken, an A indicates the highest quality. In order to get stamped with one, a rabbit has to be very clean with no signs of blood clotting caused by incomplete draining. Soaking the carcasses in salt water during cool-down helps achieve this. It should also be blemish- and bruise-free. Since dead and drained rabbits can't bruise (because there's no blood), in order to prevent bruising, it's imperative you treat your live rabbits gently and carefully before slaughter. Grade A rabbits shouldn't have any broken bones, with the exception of where the legs were clipped to remove the feet, and should be clean of any fur, bone shards, and dirt.

When grading your rabbits, look for a moderate amount of fat around the kidneys as well as around the crotch and inner walls of the body cavity. You want your rabbits to have a broad back, wide hips, and deep-fleshed shoulders. Rabbits that are thin, lean, rangy, bloody, bruised, or bony fall into B or C, depending on the degree.

Packaging

If you have a vacuum sealer, you can use that to package whole rabbits or parts. If you don't, I recommend using poultry shrink bags, which expel all the air in the bag and shrink to the meat inside when they are dipped in hot water. Poultry shrink bags are very easy to use and are available from several sources online.

While there are no crystal-clear regulations about how long you can keep rabbit fresh after slaughtering, my slaughterhouse recommends a maximum of 10 days before cooking or 7 days before freezing. In order to keep fresh, they need to be properly packaged and consistently kept at 40°F (4°C) or below. Rabbits can be held in a crystallized state in temperatures between 25 and 30°F (−4 and −1°C) for even longer without ever actually freezing. Properly packed frozen rabbit can keep indefinitely, though quality may begin to degrade after a year.

———

Compared with the more common livestock like cattle, pigs, and sheep (whose domestic roots are so ancient, they go back to cavemen days), the history of the domestic rabbit is fairly brief. It was only as recent as the Middle Ages that Europeans began to trade the first domesticated rabbits.[2] It was also during this time that this then-novel protein found its place on the French, Spanish, and Italian dinner table.[3] To date, populations in these countries still consume more rabbit meat per capita than most other places in the word.[4]

America's culinary relationship with rabbit, on the other hand, has not been so steady. Around the turn of the 20th century, rabbit was pretty widely consumed in the United States.[5] During this period rabbit had a reputation for being a sort of lowly protein, most commonly used among disenfranchised populations, such as recent immigrants and the rural poor. There was a widespread uptick in rabbit consumption during World War II, when the majority of our beef was being shipped overseas to feed the troops. Thanks to some powerful encouragement from the USDA, thousands of backyard rabbitries were established to fill the new dietary void, and Americans from all walks of life were eating the new white meat.[6] Recipes for rabbit were published in popular magazines for the very first time.

War-era copywriters at *Gourmet* even crafted a rhyme to guide read-ers through the culinary transition: "Although it isn't our usual habit / This year we're eating the Easter Rabbit."[7]

After the war ended, beef was once again readily available and rabbit lost its place on the American dinner table, falling back into obscurity—only to return roughly once a decade. In the 1960s Julia Child brought rabbit back into fashion, undoubtedly thanks to her time spent in French kitchens.[8] Then in 1985 the *Los Angeles Times* predicted a resurgence with the headline "Rabbit Renaissance."[9] Newspapers like the *Washington Post* and the *New York Times* have published articles about the rising popularity in rabbit meat every few years since. And yet today rabbit is still pretty hard to find. Perhaps the problem is that there aren't enough producers to meet demand during these surges in popularity.

CHAPTER 11

Marketing

There's a stereotype out there about the ornery, antisocial farmer who just can't deal with people long enough even to sell his potatoes. I don't know where it comes from or how long it's existed, but if there's any truth to it at all, I'll tell you it definitely doesn't apply to me. Connecting with customers is probably my favorite part of my job. I love waking up early to vend at the farmers markets all day and even hang around our CSA pickups just for fun when I'm off the clock. So when it comes to marketing, I usually take the lead. If you, on the other hand, are the farmer who spurred the creation of that stereotype, or if you just happen to be new to the marketing world, this chapter has a few tips and tricks of the trade for you.

Pricing

Because commercial production in the United States is limited, retail pricing for rabbit meat is fairly consistent, regardless of how or where it is raised. In sharp contrast with the chicken, eggs, and pork we produce on our farm, the price we need to charge per pound for our rabbit pretty much matches what is available in the grocery store. A

Letterbox Farm's stand at market. *Photo by Nichki Carangelo.*

lot of the time, we are even cheaper. This is not because we charge too little. It is because rabbit meat is fairly rare, and there simply is not a consistent, cheap source of it in the marketplace. Therefore, even big factory farms with very low costs of production are able to charge a premium for it, so they do. In our area, rabbits retail for between $7 and $15 per pound. Here, they are usually only found in specialty stores or by special order at a butcher shop. Online, both domestic and imported rabbits cost the same or more, with some retailers charging as much as $20 per pound.

We raise our rabbits in the pasture-wire hybrid system because we believe it is better for animal welfare. However, in our experience, this model does not typically fetch a higher price than conventionally cage-raised rabbits do. This may change as rabbit becomes more commonplace, but today we raise rabbits sustainably because it is better for our animals and our community—not because there is any financial incentive. Thankfully, because all rabbit garners a good price in most American markets, there is still money to made in raising them humanely like we do, even if it's not *more* money than the factory farms are getting.

Packaged rabbit ready to sell.

Common Labeling Terms

Clients and potential customers may ask you what *kind* of rabbits you have for sale. Knowing and using the common labeling terms can help ensure your markets are getting exactly what they are looking for.

Fryer or young rabbit. These terms refer to rabbits that weigh between 1½ and 3½ pounds (0.7–1.6 kg). Typically these rabbits are less than 12 weeks of age, but on our farm a 3½-pound rabbit can take as long as 16 weeks to produce. The meat on fryer rabbits is tender, fine-grained, and a pearlescent pink color. Fryers take well to almost any standard chicken preparations.

Roaster or mature rabbit. These terms can refer to adult rabbits of any weight but are most often used to label rabbits that are both over 4 pounds (1.8 kg) dressed and over eight months of age. The meat on roasters is firmer and more coarsely grained than that on younger rabbits. They also tend to be darker in color and less tender. Despite their name, roasters are best prepared using braising or stewing techniques.

Giblets. In rabbits the liver, heart, and kidneys are considered the giblets. They can be lightly fried or sautéed and used in terrines and pâté.

Head on. Thanks to a renewed appreciation for more rustic culinary traditions, many restaurants are looking for rabbits processed with the head still attached.

Tips for Approaching Restaurants

If you are new to sales, contacting potential clients can be intimidating. I remember the first time I cold-called a restaurant like it was yesterday. I was trying to sell rabbits from what was then our little backyard rabbitry, but I had no idea what I was doing. I looked up all the restaurants within a 20-mile (32 km) radius online and spent forever checking out each menu trying to guess which one could use some obscure protein on it. Finally I settled on a French bistro. I didn't know back then how common rabbits are at French dinner tables, but I figured any chef that served snails surely wouldn't bat an eye at a bunny. Next I dialed the number, it rang, and then the hostess picked up. As soon as she finished saying, "How can I help you?" I panicked and hung up. I called again, a few hours later, but this time instead of hanging up I said something like, "My name is Nichki . . . and . . . I have some . . . rabbits? For sale?" The hostess, clearly annoyed, reminded me that it was 6:00 PM and the chef was very busy with dinner service. Could I call another time? I felt so dumb.

Thankfully, every call I made after that got a little bit easier and each time, my pitch became clearer and more concise. The quicker to the point I was, the more likely I was to make a sale. I'm still not the best at working with restaurant accounts, so my business partner Faith takes on that role for our farm now. She worked in kitchens for a few years before getting started in agriculture, so she has good instincts for what chefs want to hear and, almost as important, when they want to hear it. Here are some tips she has shared with me over the years.

Do Your Research

Not every restaurant is the right fit for every farm. Before you reach out, do a little sleuthing. Is rabbit on the menu or ever listed as a special? If not, would rabbit make sense as a new addition? Look for restaurants that advertise as farm-to-table or nose-to-tail, or that serve other types of game, like quail or pheasant. These are the chefs likely to be most excited about your rabbits.

Second, look for restaurants that operate at a volume complementary to your own scale of production. High-turnover restaurants with set menus are often the most difficult for small farms to work with. They can require the exact same products in large quantities every week of the year. This is great for an experienced rabbit producer who knows they can supply X number of same-sized rabbits per week year-round, but not ideal for start-up farms that may experience product inconsistency and lapses in availability.

In our experience, working with owner-operator chefs with seasonal menus is usually the best fit. The kitchen teams in these types of restaurants are used to ever-evolving dishes and can adapt to changes in product availability more readily.

- **Call or visit during dead hours.** In restaurants that serve lunch, chefs usually arrive around 10:00 AM to prepare for service. Between 10:00 and 11:00 AM is a good window to call, but between lunch and dinner is even better. When approaching restaurants, call or stop in between 2:00 and 5:00 PM for your best chance of talking with someone in the kitchen. If it's possible, try to set up an appointment ahead of time, assuring the chef you won't need much of their time.
- **Bring a sample.** All good chefs love to play with high-quality foods. The best way to grab their attention is to bring them some beautiful sample product. Stopping by with a nice, whole fresh rabbit is sure to make their day.
- **Bring your contact information.** Though I forget to follow this rule embarrassingly often, never get caught without a business card—especially if you are a new business without name recognition. It's very easy for people interested in your products to forget your name or how to find you.
- **Have your ordering information ready.** When working with restaurants, we have found it best to be quick and to the point. Faith stops into new restaurants equipped with a laminated availability

list that includes clear information about what product is available and when. This information sheet also contains instructions for ordering, our delivery schedule, and our process for invoicing. Getting this sheet of paper into the right hands ensures that the restaurant has all the pertinent information, even if your meeting gets cut short. Laminating it makes it more likely to hang around in the kitchen, where grease and spills can result in regular paper going into the trash.

Open lines of communication. After your phone call or visit, be sure to follow up. Restaurants are stressful, busy places, and it's easy for your meeting to get lost in the shuffle. Sending an email thanking them for their time is a great way to remind them of who you are, what you have available, and how to get in touch to place an order. Our chefs increasingly contact us via text because it's quick and easy, so be sure to include a phone number at which they can reach you.

Make it easy to order. There are more and more options for restaurants to acquire high-quality, farm-fresh foods. Farmers who make it easiest to order are the ones who get the sale. It is crucial to have a clear process in place and respond to orders and inquiries promptly. While some chefs are sympathetic to how busy a farmer's day is (shout-out to ours!), many don't understand, or are equally busy, and will simply order from elsewhere if they don't hear back from you quickly.

Deliver in full and on time. This one is simple but so important that I'll say it again: Deliver in full and on time. If you can't, make sure you communicate with the restaurant as early as you can, so they have time to adapt before service.

Working Around Easter Bunny Imagery

Back in 2014 Whole Foods added rabbit to its product line for the very first time.[1] Prior to announcing this addition, the grocery store underwent what they called the Whole Foods Market Rabbit Standards

Development Process. This process mandated that all of their rabbit producers meet strict and extensive animal welfare standards, which in my opinion was pretty impressive.[2]

Despite their promise to work with only the most humane producers, small but impactful protests broke out at several locations across the country.[3] It was clear: Many Americans were not ready to eat the Easter Bunny, and boy were they vocal about it. Whole Foods pulled rabbit from its stores the following year and hasn't brought it back since.

I don't want you to take this anecdote and conclude that there is no market for rabbit. There obviously is; otherwise Whole Foods would not have added it to their product line. This is not the kind of decision major corporations tend to make lightly. Rather, I share this story in order to highlight the emotional connection many Americans have with rabbits. If you are going to raise rabbits for meat, you are bound to encounter people who will take it personally.

While rabbits were initially domesticated for the purpose of meat production, sometime during the 1700s people started breeding bigger, cuter versions to keep as household pets.[4] Shortly thereafter, the onset of industrialization encouraged many rural families to move to the city, bringing their rabbits with them.[5] It became increasingly common in these urban settings to bring rabbits out from the yard and into the house. Given this historical context, we could theorize that perhaps it was the rabbits' ability to summon fond memories of the rural past that solidified their spot in our nostalgia-prone hearts.

Soon after, rabbits began to be promoted in connection with babies and young children. Iconic books like *The Tale of Peter Rabbit* were written, and the rabbit became the bunny. For many, this romantic attitude toward rabbits persists. By the turn of the 21st century, more than two million American households had pet rabbits.[6] It's their status as companion animal that makes some people uncomfortable eating rabbits rather than their "cuteness," in my opinion. After all, have you ever seen a piglet? Or a lamb? They're pretty darn cute, yet most people have no qualms about eating them.

So, while I don't want you to be afraid, I do want you to be prepared for the occasional "How could you?" that I'm positive you will receive. When it happens, stay calm and remember there is real

169

historical context for human emotion toward rabbits. Just don't let them make you feel bad about your work. While some people really are just plain unreasonable, I have found that most often, a polite and civil conversation goes a long way into opening up minds. I also find it fairly easy to defend work that I am proud of, and I am proud of the way we raise our animals.

CHAPTER 12

The Economics of Sustainable Rabbit Production

I n the following pages I will describe the economics of the small-scale rabbitry, so that you will have all the information to ensure you are getting paid, even if farming isn't the most lucrative job out there. But first, a little background into the economics of farming in general.

Not long ago I was having a conversation with someone who told me he was looking at land to purchase and start a farm. While five years ago I might have thought of him as competition, today I'm way more concerned about the rapid loss of agricultural lands to irreparably damaging developments, so I leaned in, eager to urge him on.

"So you're considering a career change? That's so exciting!" I said, but I was thrown off guard when he responded, "No, I just want to grow something easy on the side, like garlic." Sure, garlic is low-maintenance compared with some other crops, but I definitely would not call it easy. Did this person really know what he was getting himself into?

I asked a clarifying question. "Do you want to do this as a hobby or to make money?" My inquiry was earnest. There are tons of good reasons to run a hobby farm, even if it's not economically viable, and I thought for sure that's what he was thinking about. But to my surprise, he considering it only for the money. "I'm only looking to net something like $60,000 per year, just to offset my current job."

When he asked sincerely for my opinion regarding his potential new venture, I had two choices. I could burst his bubble, or I could lie. Never one to lie about the complicated state of American agriculture, I told the truth: "I hate to be a bummer, but you should know—there is no easy money in farming." Trust me, I've looked for it. My business partners have looked for it. Virtually every farmer I know has looked for it. It sucks, but I had to tell him: Netting a $60,000 profit on any farm is pretty hard. Doing it on a start-up, with no experience, and as a side gig? Impossible. There's a reason for the old adage: "Want to have a million dollars from farming? It's simple—start with two million."

It's super easy to lose money farming, especially when you're not prepared to give it the effort or attention it requires. That's not to say you can't make a living farming. I do, my team does, and so do millions of people around the world. Rather, it's important to be aware that turning a real profit in agriculture requires significant energy, effort, and focus. Today it's a career best suited for people who like their workdays full and their bodies tired. It's a job for the researchers, problem solvers, and risk takers.

My mom always told me, "Do what you love, and the money will follow." As it turns out, she was wrong in my case, but that's okay. We definitely do not make a lot of money on our farm, but what we do make, we earn honestly and thoughtfully. We have what we need and even some extra. Plus, what our farm can't provide by way of cold, hard cash it gives in other ways. We are healthy, autonomous, and extremely well fed. It also helps that we have a million-dollar Catskill Mountain view. With all that being said, I sincerely hope there will come a day when agricultural workers are more fairly compensated for their hard work. The industry as it stands is broken, but that's a subject for another book.

The economics of your rabbit enterprise will depend on your initial investment, operating expenses, breeding schedule, sales, and marketing and overhead costs. I'll be breaking down each one of these costs using actual data from our records. Even though there is nothing particularly unusual about our farm, every business is different, and

Table 12.1. Costs for Equipment

Item	Unit	Cost
Wire cage	Each	$33.00
Feeder	Each	$5.00
Plastic tubing	Per foot (30 cm)	$0.25
Nipple waterer attachment	Each	$0.75
5-gallon (20 L) bucket	Each	$5.00
Pasture pen	Each	$300.00

prices vary from region to region. You will need to make adjustments to these figures to fit your unique situation.

Initial Investment

Your initial investment includes all of the equipment you need to purchase once, during the start-up phase of any enterprise. One of the best things about small livestock like rabbits is that they require a relatively low initial investment, one comprising light and portable infrastructure. Assuming that you have a structure like a barn, shed, garage, or greenhouse to host your rabbitry, the following sections describe all you will need to get started.

Housing and Accessories

In our system this includes all the cages, pasture pens, feeders, and waterers. Our actual costs for these items when we purchased them can be found in table 12.1.

Breeding Stock

Budget for your desired number of does plus a minimum of two bucks. If you are following our model, getting does in increments of four will reduce your per-doe cost, since each tractor can hold four litters at a time. We always try to run our livestock enterprises at max capacity—it's more bang for the same buck.

Table 12.2. Equipment Costs per Doe

Item	Cost
Baby Saver wire cage, 3 × 2½ × 1½ feet (91 × 76 × 46 cm)	$33.00
68-ounce (1.9 kg) feeder	$5.00
Plastic tubing	$0.50
Nipple waterer attachment	$0.75
5-gallon (20 L) bucket (1 per 10)	$0.50
Pasture pen (1 per 4)	$75.00
Breeding doe	$30.00
Total	**$144.75***

* For the sake of simplifying, I'll round the total number to **$145** to use moving forward.

In our neck of the woods, breeding does cost between $20 and $40, depending on the source. Some breeders charge upward of $60 per rabbit, but these are usually show-quality animals, which is not necessary for meat production. Because many of the items in the following table are used for more than one doe, it is helpful to budget using a per-doe cost. If you are maximizing your resources, and prices in your region are similar to ours, your equipment costs should look something like table 12.2.

Fixed Costs

A new rabbitry will also have fixed costs that stay the same no matter the size of your operation. At a minimum, these fixed costs include a wheelbarrow, a few buckets, a couple of transport crates, a shovel, wire brushes for cleaning cages, a pair of J-clip pliers to assemble your wire cages, some cleaning supplies, and a small rabbit wellness kit.

Using the numbers in table 12.3, we can now estimate the initial investment for a rabbitry at different scales using the equation:

$$(\text{\# of does} \times \$145) + \$400$$

Table 12.3. Fixed Costs

Item	Cost
Miscellaneous supplies	$220
Breeding buck × 2	$60
Transport crates × 2	$120
Total	**$400**

Table 12.4. Start-Up Costs by Scale

Number of Does	Initial Investment
8	$1,560
16	$2,720
28	$4,488
40	$6,200

Table 12.4 shows how it plays out. Remember, we like our rabbits in increments of four.

Operating Expenses

Your operating expenses include any costs associated with producing a product after you have made your initial investment. These expenses will change based on production and include feed, transportation to and from the slaughterhouse, processing, labor, and depreciation on your equipment.

Depreciation is the total cost of your equipment divided by the number of years before it needs to be replaced. We use 10 years as the term for depreciation in our rabbitry, so our annual expenses look something like table 12.5.

Table 12.5. Sample Annual Expense Budget

Number of Does	Initial Investment	Cost per Year
8	$1,560	$156
16	$2,720	$272
28	$4,488	$449
40	$6,200	$620

Now that we have this number, we can put an operating budget together. In order to estimate the annual cost of feed, use the formula from chapter 8:

(1,003.75 × feed price per pound) × number of does
(474.5 × feed price per kg) × number of does

The sample operating budget in table 12.6 is for a 28-doe rabbitry, producing 1,000 fryers annually. The terms used on this chart are the actuals for our business, but they will need to be adjusted to suit yours.

To calculate labor, we assume that we spend 30 minutes per day on chores plus an additional two hours per week on other tasks like breeding, loading, and regular maintenance. This came to 286 hours per year, but I rounded up to 300 here for good measure. Because our rabbits tag along with our meat birds to the slaughterhouse, we do not add in extra processing time, so your labor requirements may look different.

Breeding Schedule

Table 12.6 assumes that each doe is being bred regularly 28 days after kindling and rearing 36 kits per year. I'm copying the schedule chart from chapter 6 here to serve as a reminder how much varying schedules affect the production in a rabbitry. Just as your breeding schedule will affect your gross sales, it will also affect your costs, so again, adjust your projections according to your specific plans.

Table 12.6. Sample Operating Budget

Item	Terms	Cost
Feed	$0.20 per pound / $400 per ton	$5,621
Transportation	($990 / 3,700) × 1,008*	$250
Processing	$5 × 1,000 rabbits	$5,000
Labor	300 hours per year at $12 per hour	$3,600
Depreciation on equipment	$4,488 / 10 years	$449
Total Operating Costs		**$14,920**

* The numbers here represent (total expense / total number of animals brought to slaughter) × total number of rabbits.

Table 12.7. Sample Breeding Schedules and Their Bottom Lines

Days After Kindling	Litters per Year	Fryers per Year	Gross Income per Year per Doe
42	5	30	$750
35	5½	33	$825
28	6	36	$900
21	7	42	$1,050
14	8	48	$1,200

Sales

Now that we have done so for the cost of production, it's time to estimate sales. In doing so, think about where you are going to sell your rabbits and set your pricing wisely. The difference of $0.50 per pound is fairly small for the individual consumer but will matter

Table 12.8. Gross Income by Price per Pound

Wholesale (750 rabbits)	Retail (250 rabbits)	Gross Income
$7.50	$8.50	$25,188
$8.00	**$9.00**	**$26,812**
$8.25	$9.50	$27,828
$8.50	$9.75	$28,641

greatly in the scheme of an enterprise at scale. Again, using our 28-doe rabbitry as an example, I have created table 12.8 to illustrate what a difference a couple of quarters can make. We tend to sell three-fourths of our rabbits wholesale, so this table reflects that practice. I am using an average weight of 3¼ pounds (1.5 kg). At Letterbox Farm, we've settled on a price of $8 per pound wholesale, $9 per pound retail.

Choosing Your Markets

Sometimes the available market determines your wholesale and retail percentages for you. This is certainly true in our case, but if demand permits, make informed decisions about where you choose to sell.

Table 12.9. Potential Income Based on Outlet

Wholesale %	Retail %	Gross Income
100	0	$26,000
75	**25**	**$26,812**
50	50	$27,625
25	75	$28,437
0	100	$29,250

Table 12.9 uses our wholesale price ($8 per pound) and retail price ($9 per pound) to set the terms. At Letterbox Farm, we operate with 75 percent wholesale and 25 percent retail.

Putting the Pieces Together

To estimate your annual profit, use this simple formula: *income – expenses*. Remember the terms we have set for these calculations (of which there are many):

28-doe rabbitry
Breeding on a 28-day cycle
Producing 36 kits per doe, or 1,000 fryers total
Averaging a dress weight of 3¼ pounds (1.5 kg)
Charging $8 per pound wholesale, and $9 per pound retail
Wholesaling three-quarters and retailing one-quarter
　of the product

We have determined, based on these calculations, that the average price per fryer in this case is **$26.81**. Remember that I reached this number by finding the combined income of the wholesale and retail rabbits and dividing it by the total number of rabbits. With this in mind, we can estimate our gross income as follows:

Gross income = 1,000 rabbits × $26.81 = **$26,810**

Based on our earlier calculations, we know our operating expenses for a rabbitry this size is **$14,919**. Finally, we are ready to put the pieces together and get our potential net profit:

Net profit = $26,810 – $14,919 = **$11,891**

Dividing our total expenses by our total number of rabbits produced gives us our cost per rabbit:

$14,919 / 1,000 = **$14.90**

This number is also known as the *break-even point*. In order to turn a profit, every rabbit has to be sold for more than $14.90.

Dividing our profit by the total number of rabbits gives us our profit per rabbit:

$$\$11{,}891 \ / \ 1{,}000 = \textbf{\$11.89}$$

To determine your gross margin, use the formula: *(income − expenses) / income*. Or in our case:

(26,810 – 14,919) / 26,810 = 0.44

Margins are expressed as percentages, so our margin here is **44 percent**.

The margin for our rabbitry the year of writing this book was actually 47 percent. (For the purpose of comparison, our meat birds came in at 31 percent, our layers at 46 percent, and our pigs at a dismal 20 percent.) Due to the scale of this enterprise, rabbits make up only a small percentage of our annual income. However, if our clients could support an increase in production, rabbits would be our most profitable livestock enterprise.

Adapting for Risk

All of the previous calculations assume your rabbitry is functioning at max capacity with only a few minor hiccups in production. However, it is likely you will run into a few problems during your time as a greenhorn. The following more conservative adjustment accounts for the same exact production expenses for raising 1,000 rabbits but assumes 20 percent of the final product (200 rabbits) has gone unsold.

Gross income = 800 rabbits at $26.81 each = **$21,448**
Total operating costs = 1,000 rabbits at $14.90 = **$14,900**
Net profit = income – expenses = **$6,548**

Table 12.11. Theoretical Budget

Initial investment	$2,140 ($145 / doe + $400 fixed costs)
Operating costs	$4,470 (this includes feed, processing, transport, depreciation, and labor—30 minutes per day, year-round, at $12 per hour)
Total expenses	$6,610 ($4,395 without labor)
Income	$6,703 (250 rabbits × $26.81)
Profit	$92 (income – expenses)

Keep in mind that if you are experiencing issues with production, rather than sales, your numbers will look different. For example, if you are having difficulty getting your does bred or finding that poor mothering traits in your breeding stock are resulting in fewer viable kits, your potential gross income will drop. That means, however, that so will some of your production costs, like feed and processing expenses.

In addition to having higher margins than other livestock enterprises, rabbits are a low-risk investment. Even if you get off to a rough start as you build production skills and markets, it is not very difficult to break even in just one season. To illustrate, let's take a look at a model scenario:

You invest in a 12-doe rabbitry with two bucks. Difficulty breeding leads to two missed litters per doe for the year. Poor breeding stock selection means the does have small litters, and each produces only 25 viable kits per year.

Then 300 rabbits are grown and processed, but sales are slower than anticipated and 50 rabbits go unsold. More than $1,000 worth of inventory sits in your freezer.

Table 12.11 shows you how the numbers would break down.

Even in this scenario, you would be able to pay off your initial investment in a single year. You would also pay yourself $2,190 for your labor and the business would essentially break even. Plus, you and your friends would get to eat 50 rabbits, which is about one a week for the year—a fun silver lining.

A Case for Integration

Adding rabbits as a part of our own integrated livestock and vegetable farm made almost no change to our overhead costs. We were already paying for a website, marketing fees, cold storage, transportation to the slaughterhouse, insurance—everything we needed to produce and sell rabbits aside from enterprise-specific equipment and labor. We were also already tending livestock year-round, so little changed in terms of required labor. Therefore, producing rabbits on even a very small scale made sense for us.

But building a profitable business with rabbits *as the sole product* would look very different. Table 12.12 is by no means comprehensive, but it illustrates what the minimum income requirements could look

Table 12.12. Theoretical Business Budget

Expense	Terms	Annual Total
Rent	$250 per month	$3,000
Utilities	$150 per month	$1,800
Market stand	$190 per week at 28 weeks, including booth, gas, tolls, labor	$5,320
Web hosting	$200 per year	$200
Insurance	$1,000 per year	$1,000
Cold storage	$50 per month	$600
City deliveries	$100 per week	$5,200
Farmer's salary	$30,000 per year	$30,000
Part-time salary	$10,000 per year	$10,000
	Overhead Total	**$57,120**
Operating expenses		$71,520
	Combined Total	**$128,640**

like for a humbly successful "rabbit-only" business. To set the terms, I used averages based on our own experience in the Hudson Valley.

Based on the estimates in table 12.12, a single farmer would need to efficiently produce and sell 4,800 rabbits per year at an average price of $26.81. This is by no means impossible, but it is challenging, and finding a market for that many rabbits at that price could prove problematic. If you haven't got a market that large, consider starting a few enterprises that are smaller and more varied. Complementary product diversity allows you to sell a smaller group of clients an increased number of items.

Raising Rabbits as a Hobby

Just because something can be a commercial enterprise does not mean it *has* to be. There are plenty of reasons to raise rabbits as a hobby, and if you eat rabbit regularly, it can even save you money. On our farm, it costs less than $5 per pound to produce our rabbits. If you are doing your own slaughtering, it's closer to $3 per pound, which is much cheaper than you will ever find it at the store. Add in the benefits of reducing your own carbon footprint, improving your soils, and preserving traditional knowledge, and you've got a winning combination.

CHAPTER 13

Preparing Rabbit

We display our meats at the farmers market in a Cambro (a restaurant-grade insulated container) packed with ice, because whole chickens and pork chops are beautiful, and people buy more when they can see it without having to ask. Usually once per market someone will pick up a fresh rabbit, and I can see the wheels start spinning in their head. Based on a few years' experience, I know they are probably asking themselves these three questions:

1. Do I have time to cook a rabbit this week?
2. Will my kids or spouse freak out?
3. How do I cut this thing up?

To quell their anxiety about question three, I start by asking them if they know how to part a chicken. If the answer is yes, I say, Great! Then you can definitely part a rabbit. If they say no, my follow-up question is: Do you have a sharp knife?

How to Butcher a Rabbit

Butchering a rabbit is simple and similar to breaking down a chicken. The following is a step-by-step guide to help you through the process.

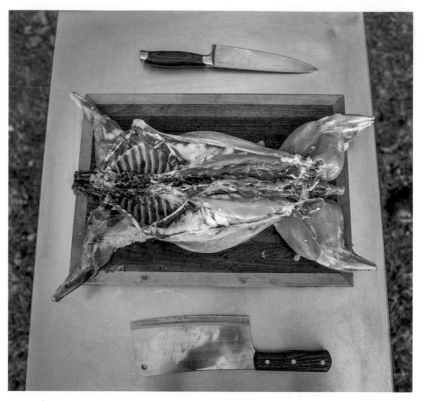

1. To butcher a rabbit, start with a sharp knife—ideally a cleaver.

2. Stand up the rabbit and use the tip of the knife to break through the breastbone.

3. Split the rib bones going down either side of the back.

4. Using the tip of the knife again, remove the hind legs at the joint.

5. Similarly, remove the forelegs.

6. Use the cleaver to break through the backbone, behind the ribs, and separate the ribs from the loin.

7. Separate the loin from the backbone.

8. When you're done, you will have eight pieces plus the backbone and the collar.

Cooking with Rabbit

Rabbits are incredibly versatile in the kitchen. With a few minor adjustments, you can basically cook them all the ways you can do a chicken: roasted, sautéed, poached, grilled, stewed, braised, shredded, and the list goes on. Laszlo and I served buttermilk-fried rabbit at our wedding.

Most cultures around the world have their own ways of cooking rabbits, and if you eat out a lot you have probably seen it on a menu without even realizing it was there. The French call it *lapin*, the Italians *coniglio*, and in Chinese it's *tù ròu*. So roast it with a *bouquet garni*, make a rustic cacciatore, poach it with soy sauce, or cook it in a tagine—it's all delicious. Just remember these three quick tips:

1. Fryers, or young rabbits that are less than 3½ pounds (1.6 kg) dressed, are the sweetest and most tender. They can be used successfully in virtually any preparation.
2. Older, larger rabbits are considered roasters or stewers. Their meat is tougher than fryers, so they should be cooked slowly— either braised or, as the name suggests, stewed. Ironically, roasters are not very good when roasted.
3. For safety, the USDA recommends cooking rabbit to an internal temperature of at least 160°F (71°C).

Those are the basics, but otherwise go nuts. If you need a place to start, here are a couple of my favorite recipes.

Hunters' Rabbit Stew

Serves 4

This recipe comes from Nonni, which is what we called my great-grandmother Maria Carlotto. It was a favorite at a yearly game dinner she hosted with my great-grandfather for a large group of friends. Rabbit was served alongside venison, quail, pheasant, chicken, and even squirrel. Great big platters of polenta made sure each one of the up-to-20 guests went home *pieni come un uovo*, or "stuffed like an egg," after a long evening of food, wine, and storytelling.

1 ounce (28 g) dried porcini mushrooms
1 rabbit, 3–4 pounds (1.4–1.8 kg), cut into pieces
Salt and pepper to taste
3 tablespoons flour
2 tablespoons olive oil
4 tablespoons butter
3 large shallots or 1 medium onion
1 carrot, chopped
2 stalks celery, including leaves, finely chopped
1 cup (235 ml) white wine
1–2 cups (235–470 ml) mushroom soaking water
3 cups (700 ml) chicken stock
1 branch (6 inches / 15 cm) fresh rosemary
4 cloves garlic
2 tablespoons chopped fresh parsley

First, soak the dried porcini mushrooms in 2 cups (470 ml) hot water. Remove the rehydrated mushrooms but save the water. Strain the reserved water through a coffee filter and set aside.

Sprinkle the rabbit with salt and let it sit at room temperature for 30 minutes. Meanwhile, dice the rehydrated porcini, and season the rabbit with pepper. Lightly dust the meat with flour. Heat the oil and butter in a large pot and brown the rabbit pieces on all sides. Removed the rabbit and set aside.

Add the shallots, carrot, and celery into the pot and cook until translucent, about 2 to 3 minutes. Deglaze these ingredients with the white wine and use a wooden spoon to scrape up any browned bits on the bottom of the pan. Reduce the wine by half, then add the reserved mushroom liquid and chicken stock to the pot. Stir to combine.

Add the rosemary, garlic, mushrooms, and rabbit all to the pot and simmer gently for 90 minutes. When it's done, the meat should be falling off the bone. Garnish with parsley and serve with polenta or a crusty loaf of bread.

Rabbit Rillettes

Makes 3–4 jars, 8 ounces (235 ml) each

This recipe comes from Geoff Lutz, who just so happens to be a very old pal and the handsome butcher featured earlier in this chapter. He often brings these tasty pâté-like spreads, called rillettes, to our big friend reunions, and everyone is always thrilled. It's a great hors d'oeuvre to make ahead of time for easy breezy hosting, not to mention a good way to introduce skeptics to the delicious world of eating rabbit.

(A note from Chef Geoff: Caramelized onions can be substituted for the leeks and capers for the green peppercorns, though it won't carry the same air of smugness that comes with the original recipe!)

1 rabbit, 3–4 pounds (1.4–1.8 kg)
1 medium onion, cut into quarters
5 bay leaves
4 tablespoons clarified butter
3 leeks, pale green and white parts cut into rings
1 tablespoon salt
2 teaspoons sugar
1 jar (110 ml) green peppercorns
Fresh parsley, finely chopped

Place the rabbit, onion, and bay leaves in a saucepan to fit. Cover by 1 inch (2.5 cm) of water. Start cooking over medium-high heat until bubbles start to form, then turn the heat down to low and simmer with the top slightly ajar until the meat is falling off the bone, which should be between 4 and 6 hours. Once the rabbit is cooked, transfer to a bowl and let rest until it's cool enough to handle. Meanwhile, strain the liquid through cheesecloth and return it to the saucepan over medium-low heat. Pick the meat from the bones and add the bones back to the liquid.

Simmer moderately while caramelizing the leeks. To do this, add the butter to a skillet over medium-low heat until it is melted. Add the leeks, salt, and sugar to pan. Cook the mixture over medium-low to low heat, stirring frequently to prevent burning, until it's soft and golden—between 30 and 45 minutes. After the leeks are done, remove them from the skillet and set aside. Remove the bones from the liquid, which should be a rich stock by now. If any debris is left behind, feel free to strain again through cheesecloth. In a food processor, blend the meat until smooth, adding stock ¼ cup (60 ml) at a time to keep it moist.

Bring the remaining stock to a hard boil, and reduce by two-thirds or until it's concentrated and can coat the back of a spoon. While the stock is boiling down to a demi-glace, fold the caramelized leeks and green peppercorns into the rabbit purée until evenly distributed. Pack the mixture into glass canning jars, leaving ½ inch (1.3 cm) or a little more space at the top. Sprinkle parsley on top, and cover with demi-glace. While still hot, place the tops on and loosely screw on collars. Refrigerate overnight, and then tighten the collars all the way. Keep refrigerated for up to 3 months, as long as the seal is not broken. To serve, spread on crackers or baguette crostini.

Conclusion

Based on our research and personal experience at Letterbox Farm, we have concluded that a pasture-based rabbitry offers good economic opportunity for the small-scale farmer, if the following criteria are being met or will be met in the near future:

Other, larger enterprises exist to help carry overhead costs. As outlined in chapter 12, in order to carry all of its own overhead costs, a rabbitry would have to scale up considerably. However, for farms that are already producing a variety of products and serving a range of outlets, rabbits can be easy to integrate. Likewise, rabbits are an asset for existing markets and enterprises, because they increase product diversity and produce free fertility for the soil.

The farm is operating year-round. An economically sound rabbitry requires efficient production. The numbers in this book are based on each doe producing six litters per year. Therefore, farms with year-round staff and sales teams will fare better than those without. However, developing markets for frozen product eliminates the need for year-round sales (although it still requires four-season production).

You can find good feed at an affordable price. There are a lot of options out there, from conventional, to organic, to growing your own. Just be sure to do the math and set your pricing accordingly.

You have consistent access to a legal processor. This is the true wild card, since it may be totally out of your control. If you live in a state that allows on-farm processing, you are all set. Otherwise contact your state health department, extension office, or department of agriculture and markets to find a licensed slaughterhouse near you.

Markets in your area support appropriate pricing and purchase in high enough volume. Selling 500 rabbits a year may seem difficult at a glance, but when you break it down, that's only 10 rabbits per week. One anchor restaurant account might be all you need to make it happen. However, it is worth feeling out your potential markets before you invest in your rabbitry. Consistent sales at a fair price are the absolute crux of your potential success.

———————

Not quite there yet? That's okay—neither were we back when we started our little broiler enterprise or our first baby vegetable garden. Starting a small rabbitry is a great way for a new farmer to get some skin in the game while they build the rest of their business. Low start-up costs along with minimal space requirements and light, portable infrastructure make rabbits an ideal starter enterprise, especially for part-time farmers without permanent land tenure.

It's also important to remember that not everything has to be about the bottom line. Whether you're raising rabbits for profit or as a hobby, for meat, fur, fertilizer, or even company, there is a lot of joy to be found in the process. I hope this guide, based on my own sometimes challenging journey, will help you to minimize obstacles and maximize this joy as you embark on yours.

ACKNOWLEDGMENTS

First and foremost, I'd like to thank my business partners, who happen to be my loving husband and my inspirational best friend. Laszlo, I would be nothing without your boundless trust and unwavering support. Thank you for encouraging me to take on this project and thank you for building me such a beautiful house to write a book in. Faith, you've helped craft the best possible version of what I once believed was just a little dream. Thank you for your foresight, intuition, and deft execution of really big ideas. Every day I look out my window and think how none of this would be possible without the two of you. What a cool little farm we have built together.

Second, thanks to every one of the team members and volunteers whose sweat has seeped into these clay loam soils. We are so fortunate that our vision was the beneficiary of your hard work. Thank you especially, Adam, Marisa, and Molly, for helping us push through a season when we had a big job to do and few resources with which to do it. Thank you, Jamie, a wonderful new addition to our livestock team. To Nina, Maggie, and Eden—I can't imagine this place without you.

To my mom, dad, parents-in-law, godparents, and Jim and Joan Gilbert—thank you for believing in us, even though we were so young and inexperienced. We never would have made it here without your encouragement, enthusiasm, wisdom, and willingness to help us slaughter chickens in a pinch.

To my brother—thanks for teaching me how to be clever and always being there when I get in over my head. Thank you, Susan Willis and Kelly and Kingsley Goddard, who showed me how to grow food when I didn't know anything. Thanks, Jim and Clint, for always processing our animals with such care.

Thank you to my photographer, Christine Ashburn, for making our farm look so darn dreamy in this book. Thanks, Geoff Lutz, for volunteering to be the handsome butcher featured in her photos.

Thank you, Mallory Murphy, whose skillful hand provided much of the reference artwork for this book. Thank you, Madeleine Bach, whose trained eye scoured my first draft for messy wording and sloppy grammar, both of which were in abundance.

Thank you Carrie Edsall, Ellen Fagan, and Daniel Salatin, who kindly opened up their rabbitries to me so I could learn a thing or two. Thanks to the folks at Northeast SARE for allowing me to learn, farm, and make a living all at the same time and for sponsoring the original guide that later became this book.

Thanks, Rob Schaffer, for all your sound advice. Thank you, Jerry Leoni, for letting me spend all day in your café while I wrote, for the meager price of a cup of coffee. Thank you to my editor, Makenna Goodman, who believed I could write a book when I most certainly did not. This has been an experience I will never forget, and I'm so honored to be a part of the Chelsea Green family.

Thank you, Matt Dzioba, in loving memory. While you never did know much about rabbits, you sure taught me a lot about life.

NOTES

Introduction

1. Joel Salatin, *Pastured Poultry Profits* (VA: Polyface, 1993).

Chapter 1: Why Rabbits?

1. Adam Starr, "Backyard Bunnies Are the New Urban Chickens," *Good*, March 4, 2010, www.good.is/articles/backyard-bunnies-are-the-new-urban-chickens.
2. Hilary Hylton, "How Rabbits Can Save the World (It Ain't Pretty)," *Time*, December 14, 2012, http://world.time.com/2012/12/14/how-rabbits-can-save-the-world-it-aint-pretty/.
3. Kim Severson, "Don't Tell the Kids," *New York Times*, March 2, 2019, https://www.nytimes.com/2010/03/03/dining/03rabbit.html.
4. Alan Farnham, "What's Up, Chef? Rabbit Is the Trendy New White Meat," *ABC News*, May 9, 2014, https://abcnews.go.com/Business/bunny-rabbit-trendy-meat/story?id=23644934.
5. Aaron Webster, "Flemish Giant Rabbits," Rabbit Breeders, August 25, 2018, http://rabbitbreeders.us/flemish-giant-rabbits.
6. Kazuko "Kay" Smith, "Raising Worms with Rabbits," Happy D Ranch, accessed April 25, 2019, http://www.happydranch.com/articles/Raising_Worms_with_Rabbits.htm.
7. Dixie Sandborn, "Bunny Honey: Using Rabbit Manure as a Fertilizer," Michigan State University Extension, Michigan State University, September 1, 2016, https://www.canr.msu.edu/news/bunny_honey_using_rabbit_manure_as_a_fertilizer.
8. Martha Wertheim, "What Are the Health Benefits of Wild Game?" Livestrong.com, https://www.livestrong.com/article/349448-what-are-the-health-benefits-of-wild-game.
9. F. Lebas et al., *The Rabbit: Husbandry, Health, and Production*, Animal Production and Health Series 21 (Rome: Food and Agriculture Organization of the United Nations, 1997), 2, http://www.fao.org/docrep/014/t1690e/t1690e.pdf.

Chapter 3: Choosing Your Production Method

1. John Wassell, "Traces of Rabbits in Harpenden and Wheathampstead," Harpenden History, Harpenden and District Local History Society, http://www.harpenden-history.org.uk/page/traces_of_rabbits_in _harpenden_and_wheathampstead.
2. Tim Sandles, "Rabbit Warrens," Legendary Dartmoor, March 21, 2016, http://www.legendarydartmoor.co.uk/rabb_warr.htm.
3. Julie Engel, "The Coney Garth: Effective Management of Rabbit Breeding Does on Pasture," Sustainable Agriculture Research and Education, December 31, 2012, https://projects.sare.org/sare_project/fnc10-824.
4. Robert Gordon Jensen, *Handbook of Milk Composition* (San Diego: Academic Press, 2008).
5. "Stress Free Chicken Tractor Plans," Store, Farm Marketing Solutions, www.farmmarketingsolutions.com.

Chapter 4: Choosing Your Breed

1. "Recognized Breeds," American Rabbits Breeders Association, https:// arba.net/recognized-breeds.
2. "American Rabbit," Livestock Conservancy, https://livestock conservancy.org/index.php/heritage/internal/american.
3. "American Chinchilla Rabbit," Livestock Conservancy, https://livestock conservancy.org/index.php/heritage/internal/americanchinchilla.
4. "Home of the Million Dollar Rabbit," Giant Chinchilla Rabbit Association, http://www.giantchinchillarabbit.com.
5. Aaron Webster, "Californian Rabbits," Rabbit Breeders, http://rabbit breeders.us/californian-rabbits.
6. "Champagne D'Argent Rabbits," Rabbit Breeders, http://rabbit breeders.us/champagne-d-argent-rabbits.
7. Information on Flemish Giant rabbits was adapted from Aaron Webster, "Flemish Giant Rabbits," Rabbit Breeders, http://rabbitbreeders.us /flemish-giant-rabbits.
8. "New Zealand Rabbits," Rabbit Breeders, http://rabbitbreeders.us /new-zealand-rabbits#breedresources.
9. "Satin Rabbits," Rabbit Breeders, August 25, 2018, http://rabbit breeders.us/satin-rabbits.
10. "Silver Fox Rabbit," Livestock Conservancy, https://livestock conservancy.org/index.php/heritage/internal/silver-fox.

Chapter 6: Breeding Basics

1. F. Lebas et al., *The Rabbit: Husbandry, Health, and Production*, Animal Production and Health Series 21 (Rome: Food and Agriculture

Organization of the United Nations, 1997), 45, http://www.fao.org /docrep/014/t1690e/t1690e.pdf.

2. Lebas et al., *The Rabbit: Husbandry, Health, and Production*, 162.

3. Lebas et al., *The Rabbit: Husbandry, Health, and Production*, 22, 54.

4. Lebas et al., *The Rabbit: Husbandry, Health, and Production*, 55.

Chapter 7: Record Keeping

1. Travis West and Lucinda Miller, "Instructions for Tattooing Rabbits," Ohio State University, https://ohioline.osu.edu/factsheet/4h-35.

Chapter 8: Feeding Your Rabbits

1. Amy E. Halls, "Caecotrophy in Rabbits," Nutrifax, January 2008, https:// pdfs.semanticscholar.org/07b5/c1ae3d1bdaf37761d1996cb81ab7c8bc 0577.pdf.

2. Halls, "Caecotrophy in Rabbits."

3. F. Lebas et al., *The Rabbit: Husbandry, Health, and Production*, Animal Production and Health Series 21 (Rome: Food and Agriculture Organization of the United Nations, 1997), 37, http://www.fao.org/docrep/014 /t1690e/t1690e.pdf.

4. Kristen Leigh Painter, "Skyrocketing Sales of Grass-Fed Beef Are Forcing the Industry to Change," *Star Tribune*, December 20, 2017, http://www .startribune.com/sustain-meat/457733503/.

5. Karen Patry, *The Rabbit Raising Problem Solver* (North Adams, MA: Storey Publishing, 2014), 108.

6. Lebas et al., *The Rabbit: Husbandry, Health, and Production*.

7. Lebas et al., *The Rabbit: Husbandry, Health, and Production*, 37.

8. Lebas et al., *The Rabbit: Husbandry, Health, and Production*, 37.

9. Lebas et al., *The Rabbit: Husbandry, Health, and Production*, 53–57.

10. V. Ravindran, *Processing of Cassava and Sweet Potatoes for Animal Feeding*, Better Farming Series 44 (Rome: Food and Agriculture Organization of the United Nations, 1995), 47, http://www.fao.org/3 /a-bp076e.pdf.

Chapter 9: Health and Disease

1. F. Lebas et al., *The Rabbit: Husbandry, Health, and Production*, Animal Production and Health Series 21 (Rome: Food and Agriculture Organization of the United Nations, 1997), 124, http://www.fao.org/docrep/014 /t1690e/t1690e.pdf.

2. Lebas et al., *The Rabbit: Husbandry, Health, and Production*, 124.

3. Lebas et al., *The Rabbit: Husbandry, Health, and Production*, 159.

4. I. Fayez, M. Marai, A. Alnaimy, and M. Habeeb, "Thermoregulation in Rabbits," in *Rabbit Production in Hot Climates* (Zaragoza, ES: CIHEAM, 1994), 33–41, http://ressources.ciheam.org/om/pdf/c08/95605277.pdf.

5. "Ammonia Monitoring in Barns Using Simple Instruments," Penn State Extension, July 12, 2016, https://extension.psu.edu/ammonia-monitoring-in-barns-using-simple-instruments.

6. Karen Patry, *The Rabbit Raising Problem Solver* (North Adams, MA: Storey Publishing, 2014), 109.

7. Diane Shivera, "Raising Rabbits on Pasture," Maine Organic Farmers and Gardeners Association, Winter 2009–10, http://www.mofga.org/Publications/The-Maine-Organic-Farmer-Gardener/Winter-2009-2010/Rabbits.

8. MediRabbit, http://www.medirabbit.com/.

9. "Domestic Rabbit Diseases and Parasites," Pacific Northwest Extension, January 2008, https://extension.oregonstate.edu/sites/default/files/documents/8426/rabbit-parasite-disease-pnw310-e.pdf.

10. For more information on ivermectin, see http://wildpro.twycrosszoo.org/S/00Chem/ChComplex/Ivermectin.htm.

11. Esther van Praag, "Protozoal Enteritis: Coccidiosis," MediRabbit, http://www.medirabbit.com/EN/GI_diseases/Protozoal_diseases/Cocc_en.htm.

12. Muhammad Fiaz Qamar et al., "Comparative Efficacy of Sulphadimidine Sodium, Toltrazuril, and Amprolium for Coccidiosis in Rabbits," *Science International (Lahore)* 25, no. 2 (2013), 295–303, https://www.researchgate.net/publication/316692292_Comparative_efficacy_of_sulphadimidine_sodium_toltrazuril_and_amprolium_for_Coccidiosis_in_Rabbits.

13. David L. Williams et al., "Eye: Conjunctivitis," Vetstream, https://www.vetstream.com/treat/lapis/freeform/eye-conjunctivitis.

14. Harry V. Thompson, *The Origin and Spread of Myxomatosis, with Particular Reference to Great Britain* (Surrey, U.K.: Infestation Control Division, Ministry of Agriculture, Fisheries, and Food, 1956), http://documents.irevues.inist.fr/bitstream/handle/2042/59488/LATERREETLAVIE_1956_3-4_137.pdf?sequence=1.

15. "Myxomatosis in the US," House Rabbit Society, July 28, 2016, https://rabbit.org/myxo.

16. "Status of Reportable Diseases in the United States," US Department of Agriculture, updated February 12, 2019, https://www.aphis.usda.gov/aphis/ourfocus/animalhealth/monitoring-and-surveillance/sa_nahss/status-reportable-disease-us.

17. Esther van Praag, "Common Fur Mites or Cheyletiellosis," MediRabbit, http://www.medirabbit.com/EN/Skin_diseases/Parasitic/furmite/fur _mite.htm.
18. Esther van Praag, "Myiasis (Botfly) in Rabbits," MediRabbit, http://www .medirabbit.com/EN/Skin_diseases/Parasitic/Cuterebra/Miyasis_botfly.htm.
19. "Reference Guides," Merck Veterinary Manual, Merck & Co., https:// www.merckvetmanual.com/special-subjects/reference-guides.
20. You may find this list in "Domestic Rabbit Diseases and Parasites," Pacific Northwest Extension, January 2008, https://extension .oregonstate.edu/sites/default/files/documents/8426/rabbit-parasite -disease-pnw310-e.pdf.
21. American Association of Veterinary Laboratory Diagnosticians, https:// www.aavld.org/accredited-laboratories.
22. Harvey Ussery, *The Small-Scale Poultry Flock* (White River Junction, VT: Chelsea Green, 2011), 219.

Chapter 10: Processing Your Rabbits

1. "Slaughtering, Cutting, and Processing," Cornell Small Farms Program, Cornell University, https://smallfarms.cornell.edu/2012/07/07 /slaughtering-cutting-and-processing.
2. Andrew Amelinckx, "A Brief History of Domesticated Rabbits," *Modern Farmer*, March 22, 2017, https://modernfarmer.com/2017/03 /brief-history-domesticated-rabbits.
3. F. Lebas et al., *The Rabbit: Husbandry, Health, and Production*, Animal Production and Health Series 21 (Rome: Food and Agriculture Organization of the United Nations, 1997), http://www.fao.org/docrep /014/t1690e/t1690e.pdf.
4. Lebas et al., *The Rabbit: Husbandry, Health, and Production*.
5. Kim Severson, "Don't Tell the Kids," *New York Times*, March 2, 2019, https://www.nytimes.com/2010/03/03/dining/03rabbit.html.
6. Karen Miltner, "Rabbit Meat Is in the Midst of a Miniature Revival," *Democrat & Chronicle*, August 29, 2014, https://www.democratand chronicle.com/story/lifestyle/rocflavors/recipes/2014/08/28/rabbit-farm -fricasee-recipe/14758685.
7. Jesse Rhodes, "Rabbit: The Other "Other White Meat," *Smithsonian Magazine*, April 22, 2011, https://www.smithsonianmag.com/arts-culture /rabbit-the-other-other-white-meat-165087427.
8. Julia Child, *Mastering the Art of French Cooking* (New York: Knopf, 1961).
9. Betsy Baldly, "Rabbit Renaissance," *Los Angeles Times*, October 31, 1985, http://articles.latimes.com/1985-10-31/food/fo-13642_1_rabbit.

Chapter 11: Marketing

1. Jenn Harris, "Whole Foods Is Selling Rabbit, and Bunny Lovers Aren't Happy," *Los Angeles Times*, August 14, 2014, Daily Dish Section, https://www.latimes.com/food/dailydish/la-dd-whole-foods-selling-rabbit-20140814-story.html.

2. "Pilot Animal Welfare Standards for Rabbits," Whole Foods Market, September 2013, http://assets.wholefoodsmarket.com/www/departments/meat/WholeFoodsMarket-PilotAnimalWelfareStandardsforRabbits-September2013.pdf.

3. Ryan Grenoble, "Whole Foods' Plan to Sell Rabbit Meat Incites Fury," *Huffington Post*, August 12, 2014, https://www.huffingtonpost.com/2014/08/13/whole-foods-rabbit-meat-protest_n_5675829.html.

4. Maya Wei-Haas, "The Odd, Tidy Story of Rabbit Domestication That Is Also Completely False," *Smithsonian Magazine*, February 14, 2018, https://www.smithsonianmag.com/science-nature/strange-tidy-story-rabbit-domestication-also-completely-false-180968168.

5. "History of Rabbits," Bunny Hugga, May 15, 2010, http://www.bunnyhugga.com/a-to-z/general/history-rabbits.html.

6. United States Department of Agriculture: Veterinary Services, *U.S. Rabbit Industry Profile*, June 2002, https://www.aphis.usda.gov/animal_health/emergingissues/downloads/RabbitReport1.pdf.

INDEX

Note: Page numbers in *italics* refer to figures, photos, and illustrations. Page numbers followed by *t* refer to tables.

ABOUT THE AUTHOR

Christine Ashburn

Nichki Carangelo is a third-generation Italian American, second-generation small business owner, and a first-generation farmer from Waterbury, Connecticut. She began her agricultural career one year after graduating from Sarah Lawrence College, and three years later she became a founding member of Letterbox Farm Collective, a cooperatively owned, diversified farm in Hudson, New York. Today she manages livestock and direct markets for the farm, while squeezing in research and organizing work on the side where she can.

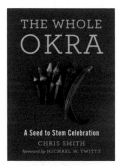

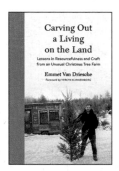
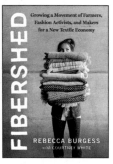